# Artificial Intelligence

*Artificial Intelligence: Applications and Innovations* is a book about the science of artificial intelligence (AI). AI is the study of the design of intelligent computational agents. This book provides a valuable resource for researchers, scientists, professionals, academicians and students dealing with the new challenges and advances in the areas of AI and innovations. This book also covers a wide range of applications of machine learning such as fire detection, structural health and pollution monitoring and control.

**Key Features**

- Provides insight into prospective research and application areas related to industry and technology

- Discusses industry-based inputs on success stories of technology adoption

- Discusses technology applications from a research perspective in the field of AI

- Provides a hands-on approach and case studies for readers of the book to practice and assimilate learning

This book is primarily aimed at graduates and post-graduates in computer science, information technology, civil engineering, electronics and electrical engineering and management.

## Chapman & Hall/Distributed Computing and Intelligent Data Analytics Series

*Series Editors:* Niranjanamurthy M. and Sudeshna Chakraborty

*Machine learning and Optimization Models for Optimization in Cloud*
Punit Gupta, Mayank Kumar Goyal, Sudeshna Chakraborty,
Ahmed A. Elngar

*Computer Applications in Engineering and Management*
Parveen Berwal, Jagjit Singh Dhatterwal, Kuldeep Singh Kaswan,
Shashi Kant

*Physics and Astrophysics: Glimpses of the Progress*
Subal Kar

*Artificial Intelligence: Applications and Innovations*
Rashmi Priyadarshini, R. M. Mehra, Amit Sehgal, Prabhu Jyot Singh

For more information about this series please visit: www.routledge.com/
Chapman–HallDistributed-Computing-and-Intelligent-Data-Analytics-
Series/book-series/DCID

# Artificial Intelligence
## Applications and Innovations

Edited by
Rashmi Priyadarshini, R M Mehra,
Amit Sehgal and Prabhu Jyot Singh

CRC Press
Taylor & Francis Group

A CHAPMAN & HALL BOOK

Cover image: Shutterstock © vs148

First edition published 2023
by CRC Press
6000 Broken Sound Parkway NW, Suite 300, Boca Raton, FL 33487-2742

and by CRC Press
4 Park Square, Milton Park, Abingdon, Oxon, OX14 4RN

*CRC Press is an imprint of Taylor & Francis Group, LLC*

*Library of Congress Cataloging-in-Publication Data*
Names: Priyadarshini, Rashmi, editor.
Title: Artificial intelligence : applications and innovations / edited by
Rashmi Priyadarshini, R M Mehra, Amit Sehgal, Prabhu Jyot Singh.
Description: First edition. | Boca Raton, FL : Chapman & Hall/CRC Press, 2023. |
Series: Chapman &Hall/CRC distributed computing and intelligent data analytics series |
Includes bibliographical references and index.
Identifiers: LCCN 2022002785 (print) | LCCN 2022002786 (ebook)
Subjects: LCSH: Artificial intelligence. | Artificial intelligence–Industrial applications.
Classification: LCC TA347.A78 A795 2023 (print) | LCC TA347.A78 (ebook) |
DDC 006.3–dc23/eng/20220407
LC record available at https://lccn.loc.gov/2022002785
LC ebook record available at https://lccn.loc.gov/2022002786

ISBN: 9781032108230 (hbk)
ISBN: 9781032305554 (pbk)
ISBN: 9781003217237 (ebk)

DOI: 10.1201/9781003217237

Typeset in Minion
by Newgen Publishing UK

# Contents

## Chapter 2 ▪ Machine Learning – Principles and Algorithms    21

Gagandeep Kaur, Satish Saini and Amit Sehgal

Arpana Mishra and Rashmi Priyadarshini

PRABHU JYOT SINGH

MELAKU NIGUS GETACHEW, RASHMI PRIYADARSHINI AND R.M. MEHRA

TOUKO TCHEUTOU STEPHANE BOREL AND RASHMI PRIYADARSHINI

# Preface

$A$ RTIFICIAL INTELLIGENCE: APPLICATIONS AND *Innovations* is a book about the science of artificial intelligence (AI)–the study of the design of intelligent computational agents.

We wrote this book because we are excited about the emerging applications and innovations of AI as an integrated science. As with any science being developed, AI applications and innovations have a coherent, formal theory and a rambunctious experimental wing. This book also covers a wide range of applications of machine learning (ML) such as fire detection, structural health and pollution monitoring and control. Deep learning (DL) plays an important role in AI. This book covers the concepts of DL algorithms of AI as well as the application of DL. The applications of DL covered in this book are aerial robotics, memoristor designing, autism and Alzheimer's disease and agriculture.

The book can be used as an introductory text on AI, ML and DL for undergraduate or graduate students in computer science or related disciplines such as computer engineering, philosophy, cognitive science or psychology.

The applications of DL and ML in different application domains will help students to understand the actual use of AI in real-life examples. While the focus is on advanced and recent applications, efforts have been made to explain the fundamental concepts in an easily understandable way with relevant references and examples.

The references given at the end of each chapter are not meant to be comprehensive lists. Rather, we have referenced the works that have been directly used and works that we think provide a good overview of the related or prerequisite literature, by referencing both classic works and more recent surveys. We hope that no researchers feel slighted by their omission, and we are happy to receive feedback if someone feels that an idea has been misattributed.

# Editor Biographies

**Dr Rashmi Priyadarshini** has a Ph.D. in Electronics and Communication Engineering from Sharda University Greater Noida. She completed an M.Tech in Electronics and Communication Engineering at IGIT, GGSIP University, Delhi, India, and a B.Tech in Electronics and Communication Engineering at Grad IETE, Institution of Electronics and Telecommunication Engineering, New Delhi, India. She is Associate Professor in the School of Engineering and Technology, Sharda University, Greater Noida, Uttar Pradesh, and senior member of IEEE. She has published 50 research papers and has been awarded "Best Teacher Award" by Sharda University. She has 15 years of teaching experience in various engineering colleges across India. She was associated with Sharda University, India, in 2006 and is currently associated with the School of Engineering & Technology as Associate Professor in the Department of Electronics & Communication Engineering. Her main research interests are design and deployment of wireless sensor networks for assistive purposes and monitoring. She has published many patents and completed government-funded projects.

**Professor R. M. Mehra** has been Professor Emeritus in the Department of Electrical and Electronics and Communication Engineering (EECE), School of Engineering and Technology, Sharda University, Greater Noida, Uttar Pradesh, India since January 2010. He received his B.Sc., M.Sc. and Ph.D. degrees in physics from the University of Delhi, Delhi, India and subsequently taught there for 38 years. He has been post-doctoral research fellow of Japan Society for Promotion of Sciences (JSPS, Japan). He was formerly HoD, Department of ECE and Dean Research, Sharda University, and is currently Advisor, Learning Resource Center and Nodal Officer, International Relations Division, Sharda University.

His primary research interests are Semiconductor Materials & Devices and Wireless Sensor Networks. He has supervised 49 PhD Research Scholars for the award of doctoral degree. He has published 215 research papers in SCI journals with ~4300 citations with *h-index 38*(SCOPUS). Currently, he is Chief Editor of *Invertis Journal of Renewable Energy*, Chief Editor, *Journal of Scientific and Technical Research* (JSTR) and member of the Advisory Editorial Board of Materials Science, Poland.

He has carried out more than 17 research projects sponsored by national (DRDO, UGC, DST, CSIR, MNRE, IUAC) and international (NSF, USA & JSPS, Japan) agencies. He has travelled widely abroad for research, teaching and consultancy assignments to universities in the USA, Japan and Europe. He is a member of several academic and professional bodies and has chaired the Semiconductor Society (India).

**Dr. Amit Sehgal** has a Ph.D. in Electronics Engineering from Dr. A. P. J. Abdul Kalam Technical University, Lucknow. He completed his M.Tech in Electronics and Communication Engineering from Guru Nanak Dev Engineering College, Ludhiana and B.E. in Electronics and Communication Engineering from C. R. State College of Engineering (now DCRUST). He is currently working as Professor, School of Engineering and Technology, Sharda University, India, and holds the additional charge of Director, Sharda Launchpad Federation. He has more than 20 years of experience in industry, academia and research. He has contributed to several research and development projects in wireless networks, health informatics, and IoT, including corporate and international academic collaboration. He is a voluntary startup mentor for Atal Innovation Mission Incubators and Atal Tinkering Labs contributing to the startup ecosystem by mentoring innovators and early age startups. Upon completing his Cambridge Certification for International Teachers and Trainers and Dale Carnegie Certificate for High Impact Teaching Skills, he contributed to the educational uplift of engineering education by serving as Master Trainer for Mission 10X–a faculty development initiative by Wipro Technologies, for more than two years.

He has written more than 40 research papers in journals and conferences and 8 patents filed. His research interests include wireless networks, remote monitoring systems and IoT. In addition, he has mentored several student entrepreneurs and has vast experience in the field of innovation and incubation.

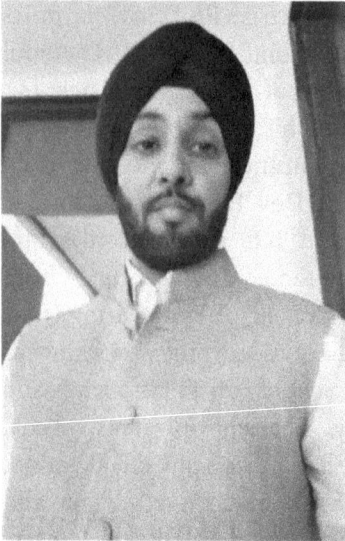

**Dr Prabhu Jyot Singh** received a B.E. in Computer Science & Engineering from the University of Rajasthan, Rajasthan, India, in 2007, and an M.Sc. in Advanced Computer Science from the University of Leicester, Leicester, UK, in 2010.

In 2012, he joined the Department of Computer Science & Engineering at Manda Institute of Technology, Bikaner, as Assistant Professor. He completed a Ph.D. in Business and Informatics from Central Queensland University, Sydney Campus in 2020. The thesis title of his Ph.D. research was "A Contribution to Communication Management in Private Unmanned Aerial Vehicle Networks". His current research interests include UAV networks, computer networks, artificial intelligence and software engineering. He is currently working as a casual lecturer in CQ University, Sydney.

# Contributors

**Melaku Nigus Getachew**
Sharda University, Greater Noida,
Uttar Pradesh, India

**Tanabalou Jayachitra**
Sharda University, Uttar
Pradesh, India

**Gagandeep Kaur**
RIMT University, Punjab, India

**R. M. Mehra**
Sharda University, Uttar
Pradesh, India

**Asha Rani Mishra**
GL Bajaj Institute of Technology
and Management Greater
Noida, India

**Arpana Mishra**
IEC college of Engineering, Greater
Noida, Uttar Paradesh, India

**OM Prakash**
Sri Venkateswara College of
Engineering and Technology,
Chittoor, India

**Rashmi Priyadarshini**
Sharda University, Uttar
Pradesh, India

**Amrita Rai**
GL Bajaj Institute of Technology
and Management Greater
Noida, India

**Satish Saini**
RIMT University, Punjab, India

**Amit Sehgal**
Sharda University, Uttar
Pradesh, India

**Prabhu Jyot Singh**
Central Queensland
University, Sydney

**Touko Tcheutou Stephane
Borel**
Sharda University, Greater Noida,
Uttar Pradesh, India

# CHAPTER 1

# Introduction to Artificial Intelligence

Gagandeep Kaur,[1] Satish Saini[2] and
Amit Sehgal[3]

[1]*RIMT University, Punjab, India, kaurgagan10deep@gmail.com*
[2]*RIMT University, Punjab, India, satishsainiece@gmail.com*
[3]*Sharda University, Uttar Pradesh, India, amitsehgal26@gmail.com*

## 1.1 HUMAN AND ARTIFICIAL INTELLIGENCE

Intelligence, in humans, is a composite of various abilities which have been considered by different groups of psychologists while defining the term. The term 'intelligence' has been derived from the Latin term *intelligentia* or *intellectus*, which is the noun form of the Latin verb *intelligere* meaning 'to comprehend' or 'to perceive'. In addition to human intelligence, scientists have been investigating intelligence or cognition in various forms of life, such as animals, birds, insects and so on. Moreover, the ability of plants to sense environmental conditions and adjust their morphology and physiology accordingly has also been explored as intelligence. The desire to embed similar capabilities in machines led to the evolution of one of the most widely explored domain of technology – 'artificial intelligence' (AI) [1, 2]. It goes beyond perceiving, understanding, reasoning, analyzing and predicting, and targets the ability to learn from experiences, adapt to changes and build new intelligent, reliable and self-healing machines and systems.

DOI: 10.1201/9781003217237-1

Though AI is considered to be the newest revolutionary field in technology evolution, the term was coined in August 1956 by John McCarthy and defined as 'the science and engineering of making intelligent machines' [3, 4, 5]. Today, applications of AI are being explored in a large variety of fields which encompasses any intellectual task, thus making it a universal field of study in the true sense. Many claims have been made that AI will replace humans. Another group of thinkers believe that it will augment human abilities and enable us to do our work in a more accurate and efficient way. Instead of being a single entity, AI has emerged as an amalgam of multiple technologies such as sensing, cognitive computing, speech, image and video recognition, machine learning (ML) and so on. The focus of developing AI systems extends to learning, reasoning, perception, problem solving and natural language communication. One of the most debated questions about the future of AI is 'whether AI can be developed to respond like a human brain.' The aim is to overcome physical limitations where a machine's cognitive abilities will be on apar with that of humans. An *ideal* performance measure for AI systems is called rationality. If a system does the 'correct thing', given what it knows, it is said to be rational. When AI reaches the point where it can behave like a human, we need to ensure that it continues to behave the same way. The performance measure regarding human-likeness of an AI entity can be based upon four methodologies [5, 6]:

- The Turing Test

- The Cognitive Modelling Approach

- The Law of Thought Approach

- The Rational Agent Approach

### 1.1.1 The Turing Test

The Turing machine, invented by Alan Turing in 1950, was used to test the ability of a machine to communicate with humans. The Turing Test, in reference to AI, suggested an operational definition of intelligence and a way to test the behavioural intelligence of machines. While undergoing this test, it is expected that the conclusion about whether the subject is interacting with a human or AI agent should become a task of utmost difficulty with minimum accuracy. The test included a human interrogator person asking questions in the form of typed messages for five minutes. If the interrogator was not able to identify whether the written responses were given by a machine or another human, the machine was considered to pass the test. An AI agent is required to possess the following qualities to achieve the goals set by the Turing Test:

- Natural Language Processing enabling the AI agent to communicate successfully in the English language (the test was designed for communication in English).

- Knowledge Representation to act as its memory and store the existing information in the form of current knowledge or new information received.

- Automated Reasoning to answer questions based on the information stored and infer new conclusions.

- Machine Learning to adapt to new circumstances by detecting patterns and extrapolating them.

A modified form of the Turing Test, called the total Turing Test, included communicating through video signals. This enabled the interrogator to test the perceptual abilities of the machine beingtestes. The additional abilities of an AI agent introduced through the total Turing Test are:

- Computer Vision for visual perception of the object.

- Robotics to handle physical objects and perform movement.

## 1.1.2 Cognitive Modelling – Thinking Humanly

Machines powered by AI agents are expected to think like a human. To make this possible, we'll need a means to figure out how people think. We must investigate the inner workings of human minds. There are three options for doing so:

Introspection: examining our ideas and constructing a model from them.

Experimental Psychology: Experimental and empirical approach on humans for their behaviour observation.

Brain Imaging: MRI-based examination of the brain functioning in different configurations and then coding it to replicate it.

When we have an adequately exact hypothesis of the human brain, it becomes conceivable to communicate the hypothesis as a computer program. The interdisciplinary field of cognitive science unites AI computer program models and psychological experiments to develop exact and verifiable speculations of the human brain. One of AI's subtler goals is to replicate the human brain's intellectual capacities using deep neural networks. The majority of contemporary neural network research attempts to mimic the inter-neuron synaptic connections in the cortex. Other human psychological abilities are becoming increasingly relevant as the brain network grows. Regardless of the challenges, we are witnessing more AI examination and execution computations which are invigorated by explicit awareness systems in the human mind.

For most, discussing the correlation between AI and human brain may appear to be a minor topic. Overall, the vast majority of people acknowledge that the most fundamental notions in AI, such as knowledge representation, reasoning and learning have evolved from the human brain's structure and functioning. However, the process through which people acquire knowledge is not fully understood. There are unique cognitive capacities that serve to supplement how knowledge is gained and formed, aside from the way knowledge manifests itself as associations between neurons. AI has begun attempting to replicate a fraction of these cognitive aspects of the human brain with the ambition of duplicating human cognition, which seems to be a never ending goal.

a. Attention is one of those enigmatic human brain abilities that we're still attempting to figure out. What brain processes cause us to focus on a single task or item while ignoring the rest of the world, and how is this attention switched from one task to another? In DL models such as convolutional neural networks (CNNs) or deep generative models, attentional systems have emerged as a new source of motivation. Recently developed CNN models, for example, have had the option of getting a schematic representation of the information and ignoring immaterial data, which has improved their ability to classify specific items in an image.

b. When we recall personal events, such as events we've attended or locations we've been, we use a function of our brain known as episodic memory. This system is frequently linked to circuits in the 'medial temporal lobe', particularly the 'hippocampus'. Inspired by the functioning of episodic memory, AI scientists have attempted to incorporate similar functionalities in the form of episodic control by implementing 'Reinforcement Learning' (RL) algorithms. Networks using RL store explicit encounters (e.g., activities and award results related with a specific game) and select new activities dependent on the closeness between the current circumstance and the past occasions stored in memory, considering the prizes or rewards related with those past occasions.

c. Continual learning is the ability of humans to learn new things while remembering past information. AI-based systems conversely experience the ill effects of what is known as the issue of 'catastrophic forgetting'. This happens, for example, when an AI agent shifts towards the optimal solution to perform the second of two progressive tasks after completing the first. Such a situation may require overwriting the solution or configuration that was used to obtain the solution of the optimal first task.

One of the new AI strategies motivated by the field of continual learning is known as 'elastic weight consolidation' used in DL techniques. This new strategy acts by reducing the learning pace in a subset of network elements recognized as important to tasks performed previously, accordingly securing these parameters to recently obtained optimal solutions. Thus, the elastic weight consolidation algorithm permits deep RL networks to enable continual learning.

d. One of the famous meanings of awareness is identified with the capacity of humans to think or imagine about what can happen in future and make plans accordingly. Most of the AI-based systems operate in highly reactive modes thus making it difficult to anticipate longer term outcomes. New AI agent techniques are being experimented with, which have the ability to produce transiently predictable sequences of samples that can give a feature corresponding to the capacity of the hippocampus.

e. The human brain's cognitive aptitude is well known for its learning abilities by inductively inferencing from previous data. Recent AI programs based upon probabilistic methods have been able to incorporate the ability to make inferences. The classes of models can make inferences about another concept and produce new sample information from a solitary sample concept.

### 1.1.3 The Laws of Thought Approach

Thinking rationally – The rules of mind are a huge set of logical principles that control how our brain works. These rules can be programmed into computer programs and used by AI systems. However, the approach to solve a problem strictly according to the laws of thought can be quite different from solving it in practice. Another important stumbling block to implementing rational thinking is the difficulty of translating informal knowledge into the formal terms required by logical notation, especially when the knowledge is unknown. The AI agent may not be able to replicate the rational thinking ability of the human brain completely.

### 1.1.4 The Rational Agent Approach

An AI agent receives precepts from the environment and decides actions to be taken back to the environment through actuators. An AI agent is rational when it acts to achieve the best outcome in the case of the deterministic environment and best expected outcome when there is uncertainty. Compared to the Laws of Thought approach, the Rational Agent approach is more general, dynamic and adaptable. A detailed description of rational agents is given in further sections of this chapter.

## 1.2 AI – AN OVERVIEW

The term artificial intelligence has been largely used to describe a branch of computer science which deals with creating computer programs to make machines think like humans and take decisions autonomously. Since the inception of this term, attempts have been made to embed various intangible features of the human brain into machines in the form of AI. These include: Reasoning, Learning, Perception from the environment, Memory, Problem Solving, Language understanding, Processing information received through other sensory organs and many more. To include these features, an integrated approach of various disciplines is required, such as:

- Neuroscience – to understand the structure and working of the human brain

- Computer Science – to develop and implement AI algorithms

- Electronics – to implement sensors and actuators

- Mathematics – to model the learning methodologies

- Biology – to understand human physiology

- Statistics – to analyze the perceived information
- Psychology – to make the machines feel and behave like humans
- Sociology – to make the machines interact like humans in the society

## 1.2.1 Goals of AI

AI is an integrated approach of various disciplines as shown in Figure 1.1. With such an integrated approach, AI systems are being developed to achieve the following goals:

- Replicate intelligence shown by the human brain
- Perform problem solving tasks by applying knowledge
- Decision making to perform actions based on percepts from the environment
- Estimate the effect of actions to be performed
- Create machines which can outperform humans in performing intelligent tasks such as:
  - Playing games like chess
  - Proving theorems of Mathematics
  - Perform clinical diagnosis and surgical operations
  - Autonomous vehicles
  - Autonomous tutor, demonstrator and adviser

## 1.2.2 Advantages of AI Systems

It's impossible to deny that technological improvements have benefitted us in our daily life. From recommending favourite music to suggesting routes to a destination, from enabling e-banking to fraud protection – these have been taken over by AI and other technology. The line separating progress and destruction is razor-thin. Every coin has two sides, and AI is no exception. AI offers various advantages which become the major source of inspiration for AI research. Various advantages of AI are as follows:

- Lessens human error
- Available 24 hours a day, 7 days a week
- Assists with repetitious tasks

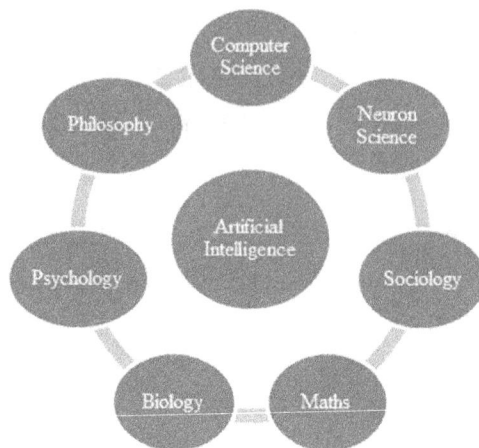

FIGURE 1.1    Various disciplines that are integrated in AI.

- Digital assistance

- More rapid decision-making

- Medical uses

- Improved security

- Efficient communication

## 1.2.3  Challenges of AI Systems

AI has had a huge impact on people's lives and the economy [7–15]. It is expected to elevate the global economy by $15.7 trillion by 2030. To put that in context, it is roughly the current combined economic output of India and China.

With several organizations projecting a 40% increase in corporate efficiency by deploying AI-based tools, the number of start-ups based on AI has seen a 14-fold increase since 2000. Applications of AI have been explored in detecting asteroids and other cosmic bodies, anticipation of diseases on Earth, and finding novel ways to combat terrorism. From design to real-time use, AI systems face many challenges. The most common challenges of AI systems are explained as follows

### 1.2.3.1  Computing Power

Most AI applications are power hungry and most of the developers find this a barrier since many of the deployment scenarios do not have the availability of such powerful computing machines. The cornerstones of AI, machine learning (ML) and DL require an ever-increasing number of cores and GPUs to perform properly. The knowledge and skills to put the learning frameworks in place across this wide range of application domains remains incomplete without the required computing power, which is a costly resource.

Although Cloud Computing systems enable engineers to work more efficiently on AI systems, they too have a cost to be paid. Not everyone can afford it, given the increased intake of massive amounts of data and the rapid expansion of complex algorithms.

### 1.2.3.2 Trust Deficit

One of the most fundamental features of AI that creates concern is the ambiguity of how learning models result in predicting the output. It is difficult for a layperson to understand how a certain set of inputs might design a solution for many types of problems.

### 1.2.3.3 Limited Knowledge

Although AI can be used as a better replacement to traditional methods in a variety of industries, knowledge about AI is the real issue. Awareness of AI's potential is limited to only a small percentage of the population, other than technology enthusiasts, technical education students and researchers.

Many small and medium-sized enterprises, for example, can manage their work in a more efficient way or learn novel techniques to enhance productivity, improve resource management, manage and sell products online, understand and learn consumer behaviour and respond to the market in a more effective and efficient way.

### 1.2.3.4 Human-level

The ability of the human brain is a critical concern for AI developers, since the expectation is to match it, if not outperform, in all cases. The applications may claim an accuracy rate of 99%, but humans can outperform them in many situations. For example, an image recognition model to determine whether the image is, for example, of a dog or a cat can perform with a remarkable accuracy of nearly 99.9%. Still, humans can virtually always predict the correct outcome, and with additional cost of infrastructure required.

A DL model would require unmatched fine tuning, hyper parameter optimization, a large dataset, and a well-defined and accurate algorithm, as well as durable processing capacity, continuous training on train data, and testing on test data to achieve equivalent outcomes. By using pre-trained models, the hard work can be avoided, but the fundamental concern is that they continue to make mistakes and would struggle to achieve human-level performance.

### 1.2.3.5 Data Privacy and Security

To train ML and DL models it is crucial to consider the availability of data and resources. Data is generated by various users throughout the world so there is a risk of being exploited.

Some innovative solutions have been developed to solve these challenges. Data is trained using smart devices, so data is not sent back to the servers; just the trained model is provided to the organization.

### 1.2.3.6 The Bias Problem

The amount of data used to train an AI system determines whether it is good or terrible. As a result, the capacity to acquire good data will be critical in the development of good AI systems in the future. The data that the organizations acquire on a daily basis, on the other hand, is weak and meaningless on its own.

They are prejudiced, and they only recognize the nature and features of a small number of people with similar interests based on religion, ethnicity, gender, community and other racial biases. True change can only be achieved by developing algorithms that can effectively track these issues.

### 1.2.3.7 Data Scarcity

Large firms such as Google, Facebook and Apple are facing allegations for illegally using user data, and nations such as India are introducing stringent IT legislation to stem the tide. As a result, these businesses are now faced with the challenge of using local data to build global applications, which may lead to prejudice.

Labelled data is used to teach robots how to learn and make predictions, which is a key aspect of AI. Some companies are experimenting with new ways, focusing on establishing AI models that can offer trustworthy findings despite a paucity of data. If information is twisted, the entire system could be poisoned.

## 1.3 AI'S HISTORY

AI is not a new term or technique for investigators. It is much older than you might believe. There are also legends of mechanical men in Greek and Egyptian myths. The historical breakthroughs in AI that characterize the path from AI generation to recent progress are discussed in greater detail below [16–19].

AI advancement (1943–1952): Warren McCulloch and Walter Pits published the first paper on AI in 1943 and suggested an artificial neuron model [20]. Donald Hebb demonstrated an improved rule for altering the strength of relation between neurons in 1949. Hebbian learning is the pseudonym given to his rule AI's Beginnings (1952–1956): 'Logic Theorist', the 'first artificial intelligence program', was created by Allen Newell and Herbert A. Simon in 1955. This program proved 38 out of 52 mathematical theorems and discovered new and more elegant proofs for a few of them as shown in Figure 1.2.

In 1956, John McCarthy, American computer scientist, created the phrase 'artificial intelligence' at the Conference of Dartmouth. For the first time, AI was recognized as an area of study. During this period, high-level computer languages such as LISP, FORTRAN and COBOL were developed. AI was a hot topic at this time.

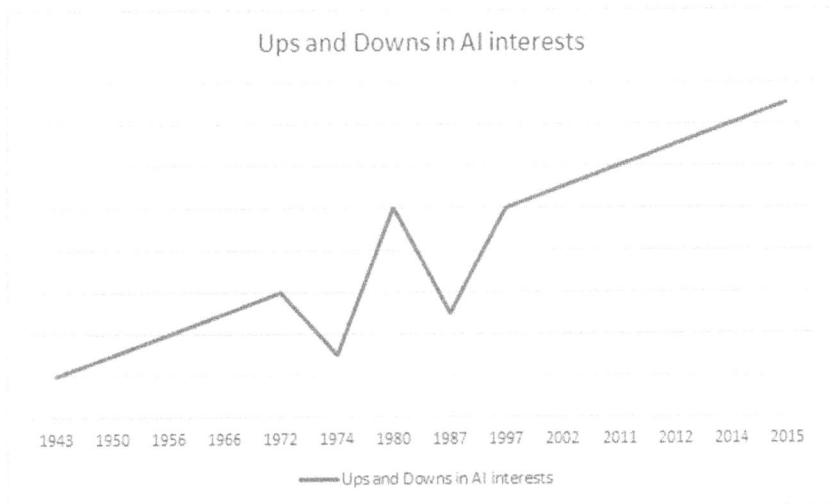

Ups and Downs in AI interests

1943  1950  1956  1966  1972  1974  1980  1987  1997  2002  2011  2012  2014  2015

▬▬Ups and Downs in AI interests

FIGURE 1.2   Yearly ups and downs in AI interests.

Early enthusiasm in the golden period (1956–1974): The focus of the researchers was on building algorithms that could solve mathematical issues. ELIZA, the first chatbot, was invented by Joseph Weizenbaum in 1966. In the year 1972, Japan produced WABOT-1, the world's first intelligent humanoid robot.

Winter in AI for the first time (1974–1980): The first AI winter was from 1974 to 1980. The period when government funding for AI research was severely limited was referred as an AI winter period. There was a dip in public interest in AI during AI winters.

AI is on the upswing (1980–1987): With the 'Expert System', AI made a comeback after a long hiatus. Expert systems are designed to emulate the decision making abilities of a human expert. In 1980, Stanford University held the first national conference of the American Association of Artificial Intelligence.

The second AI winter period (1987–1993): The second AI Winter occurred between 1987 and 1993, when investors and government once again ceased sponsoring AI research due to exorbitant costs and ineffective results. XCON, for example, was a very cost-effective expert system.

Intelligent agents emergence (1993–2011): In 1996, IBM Deep Blue defeated world chess champion Gary Kasparov, becoming the first computer to defeat a chess champion. For the first time in history, AI entered the household in the shape of a vacuum cleaner, Roomba in 2002. AI first appeared in the business world in 2006. AI is now being used by companies such as Facebook, Twitter and Netflix.

Artificial general intelligence, DL and big data (2011–present): In 2011, IBM's Watson won Jeopardy, a game show in which it had to answer complicated problems and solve riddles. Watson had demonstrated that it could comprehend plain language and solve complex problems fast. In 2012, Google released an Android app called 'Google Now', which gives users information in the form of a prediction. Chatbot 'Eugene Goostman' won a competition in the controversial 'Turing Test' in 2014. IBM's 'Project Debater' debated with two expert debaters on tough themes and did a superb job in 2018. Also in 2018, Google presented an AI application named 'Duplex', a virtual assistant that took a hairdresser appointment over the phone while the person on the other end didn't realize she was conversing with a computer.

AI has finally reached a pinnacle degree of development. At the present, big data, data science and DL are all hot topics. AI is increasingly being used by companies like Facebook, Google, IBM and Amazon to produce fantastic technology.

## 1.4 AI WORKING

Building an AI system entails reversing our characteristics and abilities in a computer and exploiting its computational power to outperform our abilities. Various sub-domains of AI are as follows:

a. Machine learning: ML is teaching a computer how to make judgements and inferences based on past experience. Without relying on human expertise to make a conclusion, it recognizes patterns, analyses previous data, and infers the meaning of the data. Firms save time and money by automating the process of drawing conclusions based on data analysis.

b. Deep learning: DL is a subset of ML. By analyzing inputs through layers, it teaches a machine to classify, infer and predict outcomes.

c. Neural networks: Human brain cells act in the same way that neural networks do. They are a collection of algorithms that record the relationship between numerous underpinning factors and analyze the data in the same way that the human brain does.

d. Natural language processing: The science of a machine reading, understanding, and interpreting a language is known as natural language processing. When a machine recognizes the user's intent, it responds appropriately.

e. Computer vision: Computer vision is basically the technique of deconstructing an image, studying various parts of the objects and interpreting using computer vision algorithms. This helps the machine learn and classify from a collection of photographs, allowing it to make better output decisions based on past observations.

f. Cognitive computing: In cognitive computing we try to mimic the human brain by analyzing images, voice, text and objects in the same way that people do, and then attempt to generate the appropriate output.

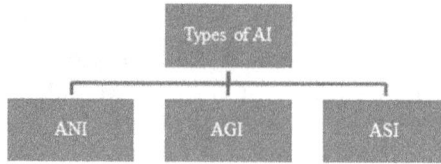

FIGURE 1.3   Types of AI.

## 1.5 TYPES OF AI

AI entities are designed for various objectives, which is why they differ. AI can be classified into three types based on functionalities as follows:

1. Artificial narrow intelligence (ANI)

2. Artificial general intelligence (AGI)

3. Artificial super intelligence (ASI)

Figure 1.3 shows the types of AI systems.

### 1.5.1 Artificial Narrow Intelligence (ANI)

These are the most popular sort of AI; they are capable of doing a single task very well and are designed to solve a specific problem. By definition, they can only do things like proposing a product to a customer or predicting the weather. At the moment, ANI is the only type of AI available. They can come close to, and even outperform, human performance in certain specific conditions, provided carefully controlled environments with a narrow range of parameters.

### 1.5.2 Artificial General Intelligence (AGI)

AGI is still a concept in progress. It's defined as AI that can do cognitive tasks at a human level in a variety of areas, such as computational reasoning, image processing, language processing and so on. We are still a long way from inventing a machine that can think itself. An AGI system will consist of numbers of ANI systems working in harmony and interacting with one another to mimic human reasoning.

### 1.5.3 Artificial Super Intelligence (ASI)

Despite the fact that we're getting close to science fiction territory, ASI is seen as the obvious next step following AGI. In every way, an ASI technology would be able to outperform humans. This would entail things like creating better art and forming emotional relationships, as well as making sensible decisions. AI systems will be able to swiftly enhance their talents and expand into worlds we could never have envisaged once AGI is attained.

## 1.6 APPLICATIONS OF AI

AI is becoming increasingly important in today's world because complicated problems can be efficiently handled in a variety of areas, including healthcare, entertainment, banking and education. Our daily lives are becoming more comfortable and efficient as a result of AI. Following are the details of some fields using the applications of AI.

### 1.6.1 AI in Astronomy

AI can be extremely helpful in resolving difficult universe problems. It can assist in gaining a better understanding of the cosmos, including how it operates, its origin and so on.

### 1.6.2 AI in Healthcare

AI has become more advantageous to the healthcare industry in the last five to ten years and is projected to have a significant impact. In the healthcare industry, AI is being utilized to produce faster and better diagnoses than humans. It can help doctors diagnose patients and keep informing doctors when a patient's condition is deteriorating, allowing medical help to be delivered before the patient is taken to the hospital.

### 1.6.3 AI in Gaming

Video games can make use of AI. AI robots can play strategic games like chess, in which the system must examine a large number of choices.

### 1.6.4 AI in Finance

The most suitable industries are AI and banking. In financial operations, ML, chatbots, automation, adaptive intelligence and algorithm trading are all used.

### 1.6.5 AI in Data Security

Cyber-attacks are on the rise in the digital era, and data security is vital for any business. AI can assist you in keeping your data safe and secure. The AEG bot and the AI2 Platform, for example, are used to better identify software defects and cyber-attacks.

### 1.6.6 AI in Social Media

In today's world, social media platforms like Facebook, Twitter, Instagram and Snapchat have billions of users and their profiles must be kept and maintained effectively. AI has the ability to organize and manage large volumes of data. AI can go through a large amount of data to find the most recent trends, hashtags and user requirements.

### 1.6.7 AI in Travel and Transport

The tourism industry is becoming utterly dependent on AI. It can help clients with a variety of travel-related chores, such as making reservations and selecting hotels, airlines and the best routes. To provide better and faster service, the travel industry is utilizing AI-powered chatbots that can communicate with clients in a human-like manner.

### 1.6.8 AI in the Automotive Industry

For better performance, certain automotive businesses are employing AI to deliver virtual assistants to its users. Tesla, for example, has released Tesla Bot, an intelligent virtual assistant. Various industries are presently working on self-driving automobiles that will make your ride safer and more secure.

### 1.6.9 AI in Robotics

In Robotics, AI plays a significant part. Typically, conventional robots are programmed to execute a repetitive task; but, using AI, we may construct intelligent robots that can perform various tasks based on their own past experiences rather than being pre-programmed by human beings. Humanoid Robots are the best instances of AI in robotics; recently, Erica and Sophia, two intelligent humanoid robots, were built that can converse and behave like people.

### 1.6.10 AI in Entertainment

We presently use AI-based applications in our daily lives with entertainment services like Netflix and Amazon. These services display software or show recommendations using ML/AI algorithms.

### 1.6.11 AI in Agriculture

Agriculture is a field that requires a range of resources to achieve the best outcomes, including work, money and time. Agriculture is becoming increasingly computerized, and AI is becoming more common in this area. Agriculture robotics, solid and crop monitoring, and predictive analysis are all examples of how AI is being applied in agriculture. AI in agriculture has the potential to be very valuable to farmers.

### 1.6.12 AI in E-commerce

AI is giving the e-commerce industry a competitive advantage, and it is increasingly demanded in the market. Shoppers can use AI to find related products in their preferred size, colour or brand.

### 1.6.13 AI in Education

AI can automate grading, freeing up more time for the instructor to teach. An AI chatbot can communicate with pupils as a teaching assistant. AI could one day function as a virtual tutor for students, available at any time and from any location.

## 1.7 FUTURE OF AI

We, as humans, have always been fascinated by science fiction and scientific achievements, and we're presently living through some of the most significant advances in human history. All across the world, various researchers, academicians and organizations are working on advances in AI and ML. AI is driving emerging technologies like big data, robotics and the Internet of Things, as well as shaping the future of human beings. As a result, skilled and qualified professionals have numerous options to pursue a successful career.

These technologies will have a greater impact on social settings and general quality of life as they progress. With advancements like AI in healthcare, face recognition, chatbots, and more, now is an excellent time to begin a career in AI. Virtual assistants have already infiltrated our daily lives, saving us time and energy. Tesla's self-driving cars have already demonstrated a big step toward the future. AI can help us predict and reduce climate change risks, which allows us to intervene before it's too late. All of these achievements are just the beginning; there is still much more to come in the fields of AI.

## REFERENCES

[1] Agrawal, A., Gans, J., & Goldfarb, A. (2017). What to expect from artificial intelligence, Magazine Spring 2017 Issue, Mitsloan.

[2] Brunette, E. S., Flemmer, R. C., & Flemmer, C. L. (2009, February). A review of artificial intelligence. In 2009 4th International Conference on Autonomous Robots and Agents (pp. 385–392). IEEE.

[3] Buchanan, B. G. (2005). A (very) brief history of artificial intelligence. *Ai Magazine*, 26(4), 53–60.

[4] Petrov, N., & Vasileva, S. (2014). History and advances of the artificial intelligence. *Science and Culture*. Taylor and Francis.

[5] Oke, S. A. (2008). A literature review on artificial intelligence. *International Journal of Information and Management Sciences*, 19(4), 535–570.

[6] Li, B. H., Hou, B. C., Yu, W. T., Lu, X. B., & Yang, C. W. (2017). Applications of artificial intelligence in intelligent manufacturing: a review. *Frontiers of Information Technology & Electronic Engineering*, 18(1), 86–96.

[7] Kayid, A. (2020). The Role of Artificial Intelligence in Future Technology. http://dx.doi.org/10.13140/RG.2.2.12799.23201.

[8] Yates, Sherry, Walker, Andrew, & Van Meter, Kerri (2020). Artificial Intelligence. www.ks.uiuc.edu/Research/namd/.

[9] Bahrammirzaee, A. (2010). A comparative survey of artificial intelligence applications in finance: artificial neural networks, expert system and hybrid intelligent systems. *Neural Computing and Applications*, 19(8), 1165–1195.

[10] Jarrahi, M. H. (2018). Artificial intelligence and the future of work: human-AI symbiosis in organizational decision making. *Business Horizons*, 61(4), 577–586.

[11] Benhamou, S. (2020). Artificial intelligence and the future of work. *Revue d'économie industrielle*, 1, 57–88.

[12] Cioffi, R., Travaglioni, M., Piscitelli, G., Petrillo, A., & De Felice, F. (2020). Artificial intelligence and machine learning applications in smart production: progress, trends, and directions. *Sustainability*, 12(2), 492.

[13] Gupta, N. (2017). A literature survey on artificial intelligence. *International Journal of Engineering Research & Technology (IJERT) ICPCN*, 5(19), 1–5.

[14] Moore, A. (2017). *Carnegie Mellon Dean of Computer Science on the Future of AI*. www.cs.cmu.edu/sites/default/files/TheLink_Summer2017.pdf.

[15] Majumdar, B., Sarode, S. C., Sarode, G. S., & Patil, S. (2018). Technology: artificial intelligence. *British Dental Journal*, 224(12), 916.

[16] Sarmah, S. S. (2019). Concept of artificial intelligence, its impact and emerging trends. *Int Res J Eng Technol*, 6, 2164–2168.

[17] Linde, H., KGaA, M., & Schweizer, I. (2019). A White Paper on the Future of Artificial Intelligence. http://dx.doi.org/10.13140/RG.2.2.32564.19844.

[18] Chrisley, R., & Begeer, S. (Eds.). (2000). *Artificial Intelligence: Critical Concepts* (Vol. 1). Taylor & Francis.

[19] Jackson, P. C. (2019). *Introduction to Artificial Intelligence*. Mineola, New York: Courier Dover Publications.

[20] Ma, L., & Sun, B. (2020). Machine learning and AI in marketing: connecting computing power to human insights. *International Journal of Research in Marketing*, 37(3), 481–504.

Web References

1. www.javatpoint.com/application-of-ai
2. www.javatpoint.com/history-of-artificial-intelligence
3. www.accenture.com/in-en/insights/artificial-intelligence-summary-index
4. www.mygreatlearning.com/blog/what-is-artificial-intelligence/
5. https://futureoflife.org/background/benefits-risks-of-artificial-intelligence/ ?cn-reloaded=1
6. https://searchenterpriseai.techtarget.com/definition/AI-Artificial-Intelligence
7. https://builtin.com/artificial-intelligence
8. www.upgrad.com/blog/top-challenges-in-artificial-intelligence/

# CHAPTER 2

# Machine Learning – Principles and Algorithms

Gagandeep Kaur,[1] Satish Saini[1] and
Amit Sehgal[2]

*[1]RIMT University, Punjab, India,*
*[2]Sharda University, Uttar Pradesh, India*
*E-mail: kaurgagan10deep@gmail.com; satishsainiece@gmail.com;*
*amitsehgal26@gmail.com*

## 2.1 INTRODUCTION

In the last few years, with the improvement of algorithms in general and the lethal development of computing power, the availability of large amounts of information has led to the increased interest and opportunities for applications of machine learning (ML) [1]. The most common deployments of ML algorithms are for classification, regression, and clustering. Another domain of application is based on quantitative reduction of large sets of data with high-dimensionality [2]. In fact, it has been proven that ML has superhuman abilities in a number of areas, such as autonomous vehicles [3], classification [4], photography, and so on. ML has also become an integral part of various daily activities through techniques such as image recognition [5], voice recognition [6], web-search [7], and fraud detection, as well as in e-mail/spam filters [8], credit scores [9], and many others with ML algorithms.

DOI: 10.1201/9781003217237-2

Although data-driven research, or rather ML, has a long history of work in biology [10] or chemistry [11], or it has only recently gained prominence in the field of science and technology.

Traditional experiments have used it to play a key role in the search for and characterization of new materials. Experimental studies with heavy resource and equipment requirement need to be carried out over a longer period of time and for several numbers or types of material. Because of these impediments, all great discoveries are mainly due to human instinct or even clairvoyance [12,13]. The first mathematical revolution in materials science was driven by the advent of computational methods [14], in particular density function theory, Monte Carlo, and molecular dynamics, which allowed scientists to explore the phase and structure of rooms much more efficiently [15]. In fact, the combined effect of both physical experiments and computer simulations can significantly reduce the time and cost of materials and structures; continuous growth of computing power [16], and development of more efficient code, should be allowed for computing and high-performance research of major device groups to display ideal experimental candidates.

Should any of these algorithms start to find their place in it, they will beat the head of the second computer revolution. Since the number of possible materials is estimated as high as in Google [17] this is definitely required. This is a paradigm shift promoted by projects such as the materials genome initiative, which aims to promote more intensive and systematic data research. A wide variety of applications in materials science, such as predicting new, stable materials [18], calculating many material properties [19], and speed of calculations based on first principles [20] have seen successful intervention of ML.

ML algorithms are already revolutionizing other technological fields such as image recognition. However, with the development of the first perceptrons, neural networks had to make a long and complex process [21,22]. To achieve significant results in materials science, it is necessary not only to play using the power of ML methods, but also to draw conclusions from other areas.

With the introduction of ML methods in the field of materials science, to date, most of the published applications are quite basic in nature and their degree of complexity. Often this involves matching models in a very small number of training sets, or even applying ML techniques to areas that can be stored in hundreds of clock cores. Of course, we can use ML techniques to provide a simple set-up process for small, low-dimensional datasets. However, this does not play to their strengths and it does not allow us to repeat the success of ML methods that we could have had in other areas.

Also, as always, when someone enters a different field of science, classification should be used correctly, for example, the term "deep learning" (DL), which is responsible for most of the past successes of ML methods (e.g., in pattern recognition and natural language processing). Of course, it's tempting to describe this work as DL. However, choosing neural networks with one or two fully connected layers and hidden learning depths may not suffice the purpose of DL algorithms [23, 24]. The success of DL is rooted in the ability of a deep neural network to teach a data format, with various levels of abstraction, without the need for human intervention. Of course, this does not apply to a neural network with two layers.

One of the main drawbacks of ML algorithms is the fact that they are often considered as a black box. Moreover, machine models can be complex, and it is difficult for people to comprehend them. We will discuss here how to use ML software and algorithms, explore them in detail, and analyze them.

## 2.2  ML APPLICATIONS

There are many ML algorithm applications. Some of the major applications are as follows:

Internet Surfing: An online search for a rating is based on what you are most likely to click on.

Computational Biology: Makes computer-aided design of rational drug based on first-hand experience.

Finance: Decides whether customer are sent credit card offers based on backend algorithms. Also includes risk assessment of credit offers and how to decide where to invest your money.

E-commerce: Provides opportunities for predicting customer churn and identifies fraudulent transactions.

Space research: Includes space research and radio astronomy.

Robotics: Tells how to deal with environmental uncertainty. Driver-less cars is an example.

Extracting information: Includes posing queries to gather relevant information using databases over the internet.

Social networking: ML provides data extracts based on relationship preferences.

Debugging: Used for error detection of computer related problems. It is a time-consuming process. It identifies where the bugs are.

## 2.3 ML KEY ELEMENTS

A large number of ML algorithms have been developed and new ones are frequently being introduced. Each ML algorithm consists of three elements:

Representation: Knowledge representation is a major key element. Decision trees, support vector machines, graphical models and others are key examples.

Evaluation: The process of evaluating the program's performance (hypotheses accuracy, predictions and opinions probability, margin, posterior probability and others are key examples.

Optimization: Tells how to create a candidate program known as a search criterion. For example, convex constrained optimization, combinatorial optimization.

## 2.4 TYPES OF LEARNING

There are three types of ML:

Supervised learning: The training data includes the desired results. This is spam, this is not a guide.

Unsupervised learning: The training data does not include the desired results. Clustering is an example of unsupervised learning. It is very hard to differentiate between good and bad learning.

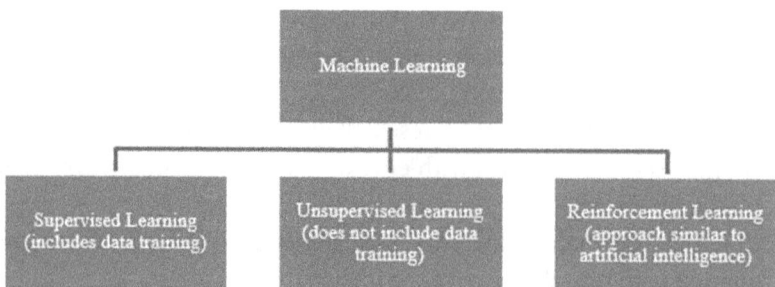

FIGURE 2.1   Types of machine learning.

Reinforcement learning: ML models are trained to produce required decision sequences.

Figure 2.1 shows the types of Machine Learning.

### 2.4.1 Supervised Learning

The supervised learning method is applied to several ML users. Supervised learning is so called because this is how the system participates in the learning process, as is done in the training dataset. And while a set of training data is being used, this process can be considered as a teacher's control over the learning process. The correct answer is widely known and stored in the system. The system will help you make predictions based on data collected during the process, and corrections are made by the teacher themselves. This is the ultimate goal of training, only if the algorithm has achieved a sufficient degree or level of performance as shown in Figure 2.2.

There are two types of supervised learning tasks. These tasks can be divided into classifications and regressions.

Classification problems: Problems whose output variables come under a specific category fall under classification problems. Say, for example, "BLACK" or "WHITE". Black or white are specific categories and the output variables has no moderate path to choose in between.

Regression: Real or true value comes under the category of regression problems. For example "DOLLAR".

There are many problems that can be observed in the data model: common problems that may arise or should be built at the forefront of classification

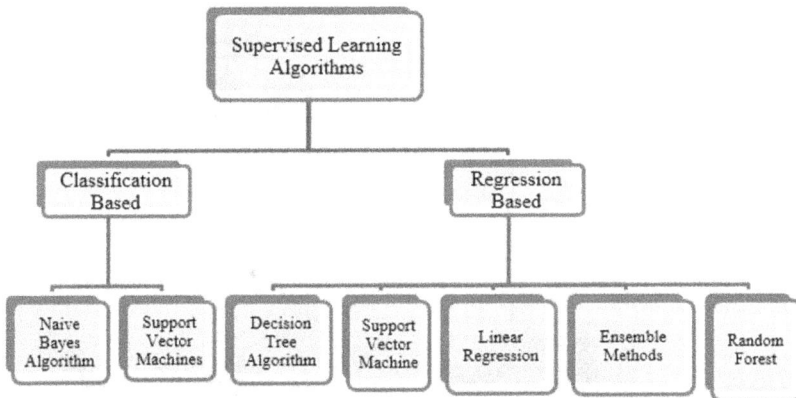

FIGURE 2.2   Supervised machine learning algorithms classification.

problems and regression problems. The most common include, among others, time series forecasting and recommendations.

There are various popular supervised ML algorithms, as outlined below.

### 2.4.1.1 Decision Tree Algorithm

Simply put, a decision tree provides support for deciding which data item to use – for example, if you are a banker, and you get the opportunity to view a decision on whether you should grant a loan to an individual. In this case, factors like profession, age and level of education matter a lot in taking the right decision. This can be done using a decision tree algorithm. Although this is perceived as a decision tree, you need to start the process starting at the root node, then go to the answer at each node and take the branch that corresponds to the specific answer. After that, you need to traverse from the root node, and then on the sheet, and then, as queries in the context of the data element. Let's take an example of starting a project based on the decision tree, as shown in Figure 2.3.

1. Write down the decision to be taken.

   To start the tree, draw a square (root node) 1for your paper. Write down the most important decisions and search results.

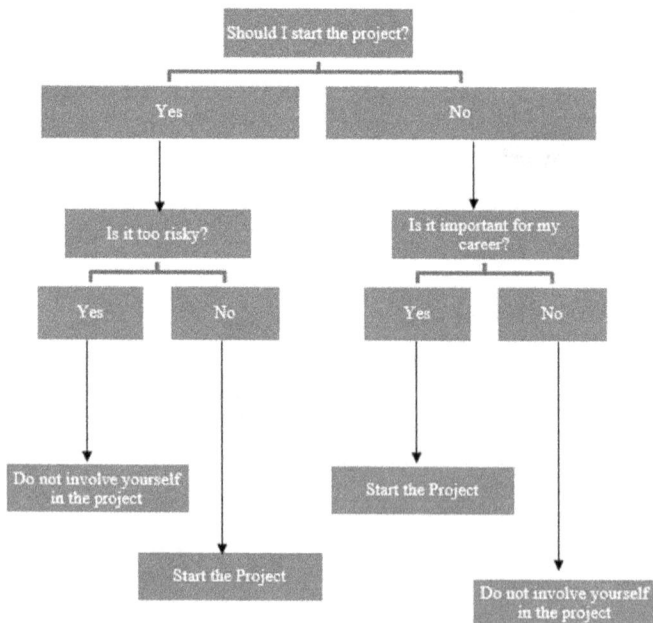

FIGURE 2.3    Example of a decision tree algorithm.

2. Drawing a line.

Draw a line leading to input fields for each possible decision or action – at least 2, but not more than 4 lines. Keep the line as far apart as possible to later place in the tree.

3. Show the results of the solution at the end of the line.

Tip: it is good practice to draw a circle where the result is uncertain and to draw a square on the consequences leading to the following problem.

4. Continue inserting boxes and lines.

Continue this process until all problems disappear and everyone is on the line, or an undefined result, or even at the end.

5. Exit the tree.

The boxes that do not define the results remain the same as they are now.

### 2.4.1.2 Naive Bayes Algorithm

This is one of the classification methods, depending on Bayes theorem. This is based on the Presumption of Predictor Independence. Simply put, this could be constructed as a naive Bayesian group, which assumes that a particular feature class is not directly related to any other features.

Looking at the example, fruit can be considered as an apple – not only based on its color, that is, if the color is red, if it has a round shape, and is approximately 3 cm in diameter. Even if these functions depend on each other, and each of the functions is available on the other, all these skills that contribute, regardless of the probability that the result is a fruit, it is like an apple, and therefore it would not be so naive.

The naive Bayesian model is not difficult to build, and it is really useful for very large data sets. Along with its simplicity, Naive Bayes is considered superior to other more complex methods and classification.

Conditional probability can be calculated using the formula given below:

$$P(H|E) = \frac{P(E|H) * P(H)}{P(E)} \tag{2.1}$$

Here, P(H) is the probability that hypothesis H is true. It is also called prior or supreme probability

TABLE 2.1   Training Data Set for Fruit Prediction

| Tenderness | Size | Color | Fruit |
|---|---|---|---|
| Hard | Small | Red | Apple |
| Hard | Small | Green | Apple |
| Hard | Medium | Red | Apple |
| Hard | Medium | Green | Apple |
| Soft | Small | Red | Tomato |
| Soft | Small | Red | Tomato |
| Soft | Medium | Green | Tomato |
| Soft | Medium | Green | Tomato |

TABLE 2.2   Test Observation Data Set for Fruit Prediction

| Tenderness | Size | Color | Fruit |
|---|---|---|---|
| Hard | Medium | Green | ?? |

P(E) is the probability of the evidence and it does not depend upon the hypothesis

P(E|H) is the probability of the evidence that hypothesis H is true

P(H|E) is the probability of the hypothesis that the evidence is there

Let's understand the algorithm with one example to predict the fruit (whether it is an apple or a tomato). Let's assume the three predictors: tenderness, size, and color. The training data set for the same is given in Table 2.1. We'll use naive Bayes to predict the fruit for the following test observation.

Now, we have to identify which posterior is greater; apple or tomato. For further classification, the posteriors are calculated as follows:

$$\text{Posterior (Apple)} = (P(\text{Apple}) \times P(\text{Hard / Apple}) \times P(\text{Medium /}$$
$$\text{Apple}) \times P(\text{Green / Apple})) / \text{evidence}$$

$$\text{Posterior (Tomato)} = (P(\text{Tomato}) \times P(\text{Hard / Tomato}) \times P(\text{Medium /}$$
$$\text{Tomato}) \times P(\text{Green / Tomato})) / \text{evidence}$$

And,

$$\text{evidence} = [(P(\text{Apple}) \times P(\text{Hard / Apple}) \times P(\text{Medium / Apple}) \times$$
$$P(\text{Green / Apple}))] + [(P(\text{Tomato}) \times P(\text{Hard / Tomato}) \times$$
$$P(\text{Medium / Tomato}) \times P(\text{Green / Tomato}))]$$

Therefore, after mathematical calculations,

$$P(Apple) = 0.5$$

$$P(Tomato) = 0.5$$

$$P(Hard/\ Apple) = 0.5$$

$$P(Medium/\ Apple) = 0.25$$

$$P(Green/\ Apple) = 0.25$$

$$P(Hard/\ Tomato) = 0.00$$

$$P(Medium/\ Tomato) = 0.25$$

$$P(Green/\ Tomato) = 0.25$$

Therefore, evidence = 0.015 + 0.00 = 0.015

$$Posterior\ (Apple) = 1$$

$$Posterior\ (Tomato) = 0$$

The posterior is greater in the case of apple. Therefore, we predict the sample as apple.

### 2.4.1.3 Support Vector Machines

The support vector machine (SVM) has become a new method of ML. It is believed that it can be used to solve both classification and regression problems. We can construct any observation as a point in an n-dimensional space (where n is the number of elements in the data set). The major task is to identify the optimal hyperplane to be successful, and to classify the data points by their respective classes.

The hyperplane is a decision boundary, which is a clear distinction between two classes, as in SVM. The data point will go both ways from the hyperplane, which can be assigned to different classes. The size of the hyperplane depends on the number of input objects in the data set. If there are 2 inputs with a hyperplane, it will be a line. However, if the number of elements is 3, then it will be a two-dimensional plane.

The reference vectors are the points closest to the hyperplane, as well as the influence of the position and orientation of the hyperplane. The maximum distance should be selected as a hyper-plane. Even a small perturbation to the vector position can change the hyperplane.

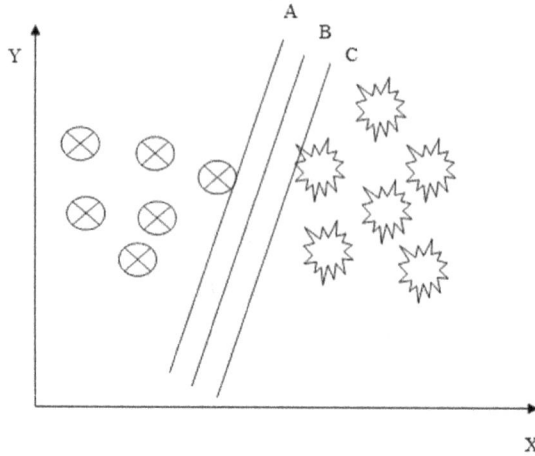

FIGURE 2.4   Two-dimensional graph to predict hyper-plane.

As we are now clear with the terminology of SVM, let's take one example where we have two different types of data as shown in the 2-dimensional graph in Figure 2.4. In this problem, line B best defines the hyper-plane because it defines the best separation between both the data types, whereas line A and line C do not support best separation.

In the case of 3-dimentional or complex data, the algorithm projects the given data into higher dimentional spaces in order to define the best separation or best hyper-plane. For projection into the higher dimension, there is a need to create a new dimension, Z.

$$Z = X^2 + Y^2 \tag{2.2}$$

Figures 2.5 and 2.6 best define the hyper-plane by projecting the data into higher dimensional spaces.

### 2.4.1.4 *Random Forest Algorithm*

A random forest is a collection of decision trees. Decision trees are grown very deeply, often devoting themselves to over-fitting information preparation so that they can produce a large variety, even small changes to the input data.

They are sensitive to specific assignments, such that they can be trained to a limited level keeping them error prone and thus not suitable for testing purposes. This algorithm helps to increase the number of such trees, and

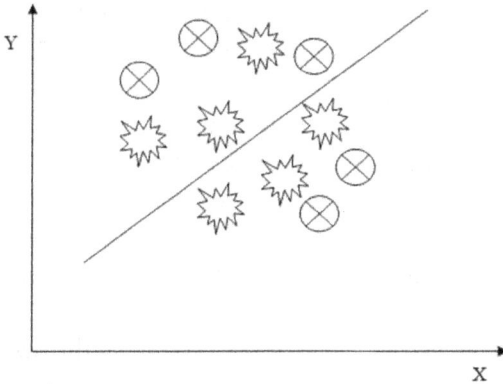

FIGURE 2.5    Complex data set.

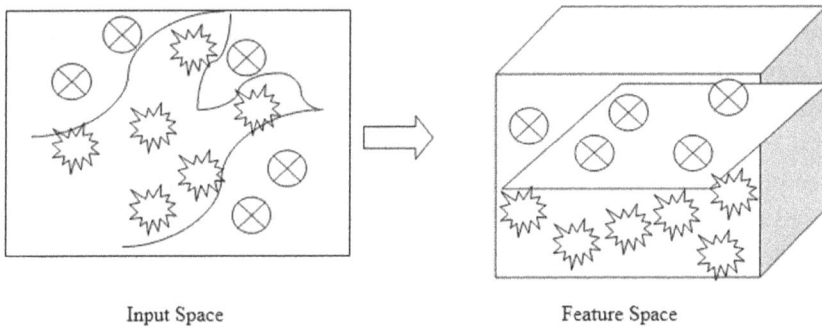

Input Space                         Feature Space

FIGURE 2.6    Input space and feature space for complex data.

decides to give the average value of another tree, classification (or location). This will reduce the variance. Trees with different formats are learned based on different components of data set preparation. From an input vector, a new object is classified by moving the input vector down for each tree in the forest. Each tree given a forest classification selects the classification that has the most votes, or the average of all the trees in the forest.

One of the popular ML algorithms that can be used for both ML classification and ML regression problems is random forest. It is based on the concept of ensemble learning. The technique includes a group containing a number of trees to produce different lower-order information sets, and an average accuracy prediction of the data set is required. Instead of relying on a decision tree, a random forest takes a prediction from each tree and, based on the majority vote predictions, predicts the final result.

Increasing the number of trees in the forest leads to increased accuracy, as well as preventing over fitting problems. Figures 2.7 and 2.8 explains how the random forest algorithm works.

Because a random forest combines multiple trees in the ability to antici-pate the output of a dataset, it is possible that some of the decision trees are able to anticipate the corresponding output port, while others are not. But, taken together, all trees have the ability to provide an appropriate output. Therefore, reasons for the improved random forest classifier are given below:

1. There must be no real value of the variable in the data set so that the group can accurately tell the result without thinking about the result.

2. Prediction of each tree, and must have a very low correlation.

For example, suppose there is a data set that includes several different items in a photo. So, this dataset is exactly for this purpose to use a group of random forests. This information is then divided into groups, and we take into account each tree and solution. At the training stage, each decision tree predicts the results, and when a new data point appears, rather than based on most of the results, the random forest classifier provides this final decision.

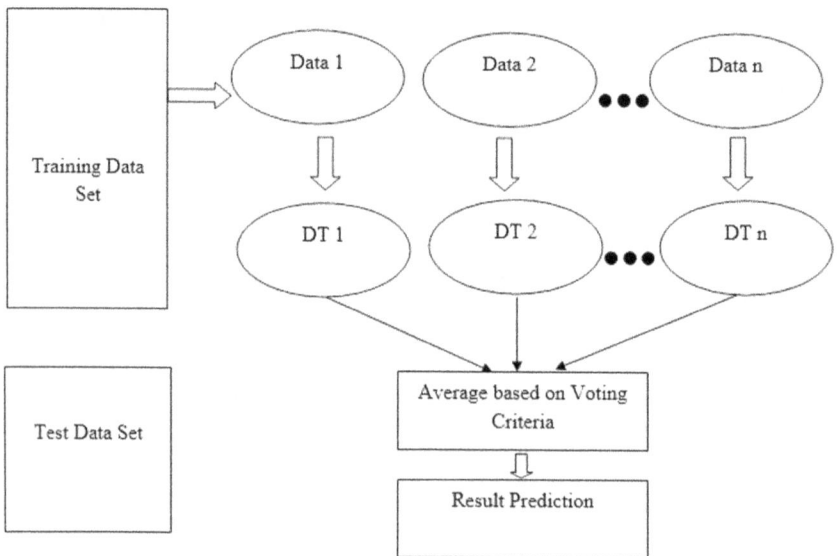

FIGURE 2.7    Working of random forest algorithm.

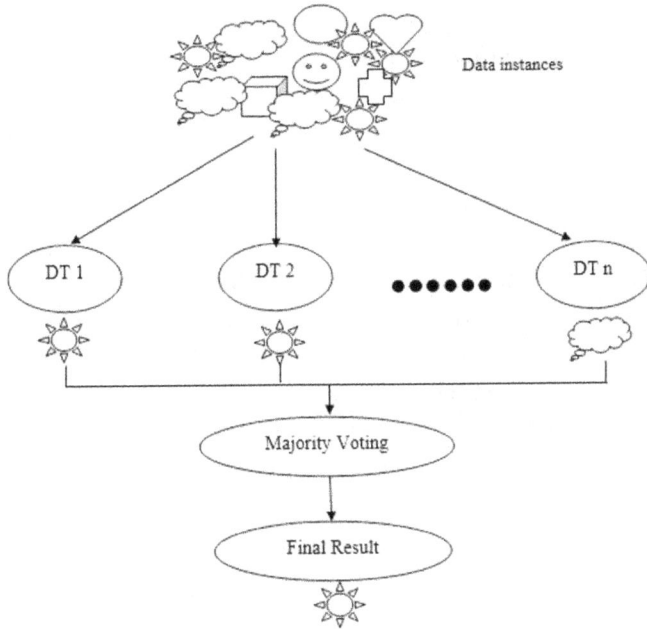

FIGURE 2.8 Example of random forest algorithm.

### 2.4.1.5 Linear Regression

As the name suggests, this is all that is well known in linear regression: it is an approach to modeling the relationship between the dependent variable y and one or more independent variables, which can be called "x" and expressed as linear. The word linear indicates that the dependent variable is proportional to the independent. There are other things to keep in mind.

"y" changes linearly as there is an increase or decrease in the value of "x". Therefore, the relationship remains constant. It can be expressed in mathematical form as follows:

$$y=Ax+B \tag{2.3}$$

Assume that A and B are to be treated as fixed values. The goal of latent, supervised learning using linear regression is to find the exact value of the constants "A" and "B", using datasets. Then the values of constants that will be useful in predicting the value of "y" in the future. Such cases, with one independent variable, are called simple linear regression. Multiple linear regression is the one which has a possibility of having multiple independent variables.

Let's say that we have a good amount of historical data that should have input x and output values/result "y"

We also know that the change in x-effects of y is somewhat linear. This means that if we have the values "x" and "y", and plot them on a graph, then we get such a graph as shown in Figure 2.9. Not exactly the same, but it's something like this, in which, supposedly, the points go in a linear direction.

The values of "A" and "B" can be obtained by looking at historical data. We can then predict the value of y for any value of x and y.

As evident from the graph, the points, though forming a linear pattern, are not on the same line. We cannot say that it fits 100% the value of "A" and "B" but what can predict the most suitable straight line (line AB).

This means that the results are approximate. There may be a slight deviation in the actual value. At the same time, it works perfectly for most business systems.

Looking at a point on the chart, what happens if you work in a single line, the distance to which each point on the graph is optimal/minimum. This will mean that "line of good match" indicates line AB in Figure 2.9.

Therefore, we draw any straight line so that the graph of the function can have any value of "A" and "B". Suppose "A" and "B" are 1 and we pull up to a line on the graph, for each of them. Based on the x values, the line can be in one of the following positions:

a. On the left side of the spot for the y-axis. Also, vertically.

b. On the right side of the spot on the x axis. In a more general form.

c. Somewhere between the points, but for a better match.

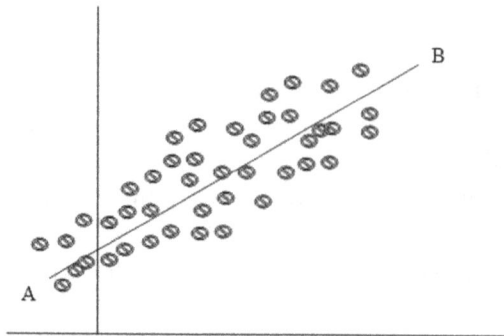

FIGURE 2.9   Linear distribution of data.

Because it was a random line, we extrapolate this line, iteratively and slowly, in the direction of the location that best fits the data and the sample size (points shown in Figure 2.9).

So, actually, we need

a. To understand whether it is a line that fits best or not

b. If it is a line of best matching, then move it up the line of best matching. This means that we need to change the value of either "A" or "B".

c. How big is the value, and do they need to be changed and in what direction? To achieve these goals, we will use a combination of gradient descent using the least squares method.

### 2.4.1.6 Ordinary Least Squares Regression Algorithm

If we look at the statistics, it is a method by which we can make an estimate of the unknown parameters. This is known as the linear regression model; it is equipped with the goal that minimizes the differences between the observed responses in the random data set.

In addition, this would minimize reactions that were very well predicted by linear approximation of the data. The resulting estimate can be expressed as a simple drawing, especially when it is found in the case of a single-use case. The value of the estimators is always consistent. On the other hand, regressors have exogenous behavior. There exists no multicollinearity property. For linear un-biased estimators, it remains optimal. Even in the case of errors, these are homoscedastic and consistently uncorrelated. Under these conditions, there is a method that provides minimal variance, mean-unbiased estimation, and errors that have finite variances. Implementing these additional assumptions, there are errors that can be well distributed. The algorithm provides a maximum likelihood estimator. The ordinary least squares regression algorithm is most commonly applied in fields of political science, engineering, and as well as having many other applications.

### 2.4.1.7 Logistic Regression

Logistic regression is a supervised ML algorithm that is used for classification purposes. However, the power in the name of one way or another is misleading, even if it doesn't fall into the trap that is a kind of regression algorithms. The name of logistic regression is a special function – a logistic function that plays a key role in this approach.

In logistic regression, the model will be called a probabilistic model. This helps you determine the probability that an instance belongs to a particular class, since this is a probability, and the output is between 0 and 1. Where we use logistic regression as a binary classifier (classifying two classes, we can consider lessons for a good class and a bad class. If we get the chance. The higher the probability that it is greater than 0.5), the more likely it is that it is in a good class. Also, if the probability is small (less than 0.5), we can assign it to the bad class.

Take the e-mail spam and ham (non-spam) formats as an example. We assume that malicious spam, which would have been in the good class, and good-natured ham, were in the bad class. So, what we can do at the beginning is use a few selected examples to email and use it to train the model. After training, this can be used to predict the class new samples based on the email. If we use the example of this model, it will give us a value of, say, y, such that $0 \leq y \leq 1$. For example, the value we get is 0.8. Using this value, we can say or even assume that there is an 80% chance that the samples were checked in by some spambots. In this regard, it can be classified as spam by email.

### 2.4.1.8 Ensemble Methods

Ensemble methods are meta-algorithms that combine multiple algorithms and ML techniques into a single predictive model to reduce variance (batching), shift (zooming in), or improve prediction (stacking). We can divide ensemble methods into two groups.

a. Methods that are not related to the group were completely removed from the student database. This number is generated sequentially (e.g., AdaBoost). The motivation for applying subsequent methods is mainly to take advantage of the dependency, i.e., between the main students. Overall performance can be increased and enhanced by looking at all examples with high weights.

b. Parallel ensemble method is a basis that students are created in parallel (e.g., a random forest). Then there is a fundamental motivation, called the parallel methods involved in using this independence, which is at the heart of students, since mistakes can be significantly reduced by learning.

Most team methods use a single basic learning algorithm to produce homogeneous base students, i.e., students who fall into the same class, which in turn results in a single team. There are a number of methods that are constantly encountered by a variety of students, that is, students of different types, resulting in a heterogeneous syndrome. For the group method to be more accurate than each of the individual members, choose from the group and students as the most accurate, and as diverse as possible.

## 2.4.2 Unsupervised Learning

Algorithms in which you don't need to load/transform the input data (X), and there is no need to specify corresponding output variables, are referred to as unsupervised algorithms. The main purpose of unsupervised learning is to model the underlying structure or to distribute data in order to help students learn more about the data.

They are called unsupervised because, unlike supervised learning, there are no correct answers, and there is no training environment involved. Algorithms are on their own to help you discover an interesting data structure. These algorithms are further classified into clustering and association.

Clustering: To find out a solution based on the grouping behavior comes under the category of clustering. Data is divided into different clusters (distinct groups) and outputs are predicted using algorithms.

Association: The principle of association is to study problems. This is where you will find the strict rules that describe most of your personal data. For example: people who buy X also tend to buy Y.

Some of the popular unsupervised learning algorithms are explained as follows.

### 2.4.2.1 K-means for Clustering Algorithm

K-means is one of the simplest unsupervised learning algorithms that solves the most well-known cluster problem. They can be grouped as those that must follow a simple and convenient way to classify a given data set using multiple clusters (assume k clusters). The basic idea is to define k-centers i.e., one for each cluster. These centers must be designed and installed in an

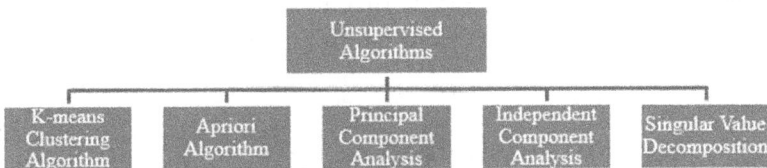

FIGURE 2.10   Unsupervised learning algorithm types.

absolutely clever way, as they have different locations, which leads to producing different results. This, therefore, may be the best choice i.e., their distance should be as maximum as possible. The next step is to take each item belonging to a specific dataset, and can be associated with the nearest center as shown in Figure 2.10. When it doesn't make sense to wait for the first step to have already been completed and ready for an early age and group, this is the part where the need for re-calculation arises. Here k is a completely new center termed as "centroid", like the center of gravity of the clusters. It comes as a result of the calculations in the earlier steps. After we get these k new weights, we need to make new bindings. It should lie inbetween the same data points and its value should be close to the previous centroid. An iterative path has to be created. As a result of this cycle, we can note that the centers will gradually change their location. This will continue until there is no further change of location, or in other words, we can say that the centers are not moving. After all, this algorithm aims to minimize the objective function, which is known as the squared error function. Following are the steps to solve k-means clustering iterations:

Step1: Choose k no. of clusters

Step2: Mark centroids. For this, you have to select any k random points from the given data set

Step3: Associate the data points from the given data set with the closet centroid

Step4: Mark centroids for the newly formed clusters

Step5: Repeat step3 and step4 until

    i.   Newly formed centroids do not change location

    ii.  Data points tend to remain in the same cluster i.e., data points do not change cluster

    iii. Iteration limit has reached its maximum

The maximum possible limit for the number of clusters is equal to the number of instances or observation in the given data set. It should not be more that the given set data points. After all, this algorithm has the goal of minimizing the objective function, referred to as squared error function given by:

$$J(v) = \sum_{i=1}^{C} \sum_{j=1}^{Ci} \left( \|xi - vj\| \right)^2 \qquad (2.4)$$

Here,

||xi-vj|| is the Euclidean distance between xi and vj
ci is the no. of data points in the ith cluster
c is the no. of cluster centers
X={x1, x2, ......xn} is the set of data or observation points
V={v1, v2,...... .vn} is the set of centers

Let's understand the algorithm using the following example. Suppose we have a given set of observation points (as shown in Figure 2.11).

By observation, it is clear that we have three different types of data points. We will now apply the k-means clustering algorithm and randomly divide the given points into clusters. Here, we have divided the given set into 3 clusters. There is no such criteria for the number of clusters. As we have three different shaped data points, so, we divided into three clusters as shown in Figure 2.12.

Now, redefine the centroids based on observational point distances and redefine the clusters. Here, we have taken a very simple example to understand the logic. After redefining and re-location, the algorithm results in different clusters as shown in Figure 2.13.

The number of iterations will depend upon the observation data types as well as the number of observation points.

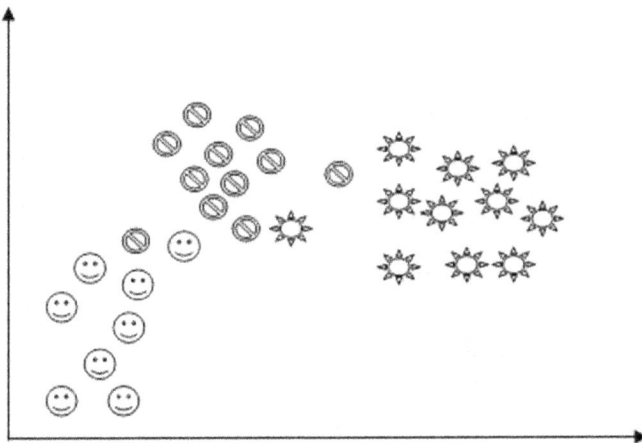

FIGURE 2.11   Observation points in a given data set before implementing k-means algorithm.

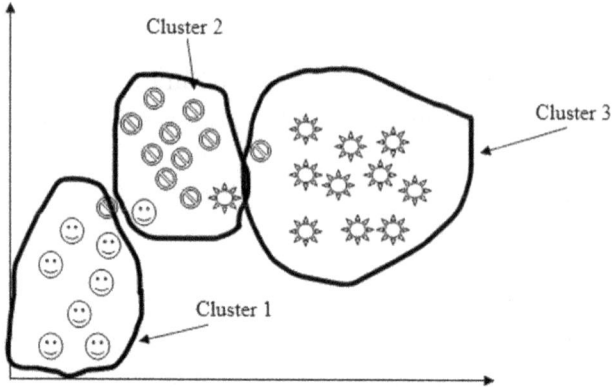

FIGURE 2.12  Result after implementing iteration 1 of k-means clustering algorithm.

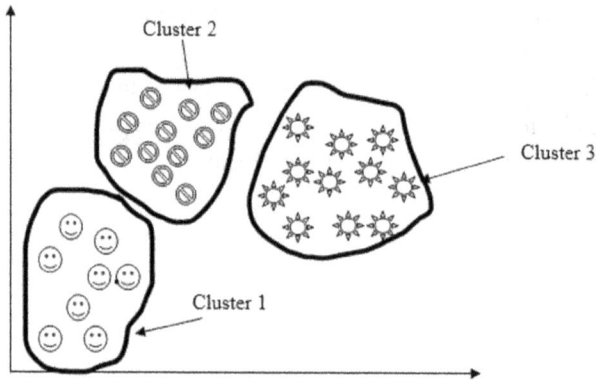

FIGURE 2.13  Result after implementing final iteration of k-means clustering algorithm.

### 2.4.2.2 Apriori Algorithm

Apriori is one of the algorithms that are often used for sample-based analysis. Often, sample-based mining techniques are employed in order to calculate the probability of occurrence of each of them based on samples. After threshold conduct, these are used in anticipation of an important assumption. For example, let's say we moved to an e-commerce website and started shopping. After that, we take our products to the cart, we may have seen a notification or sidebar, which says that "People who bought these products also got this". This conclusion is probably one of the most common sample-based mining algorithms. These algorithms scan customer transactions and find purchases of one or more products to find out

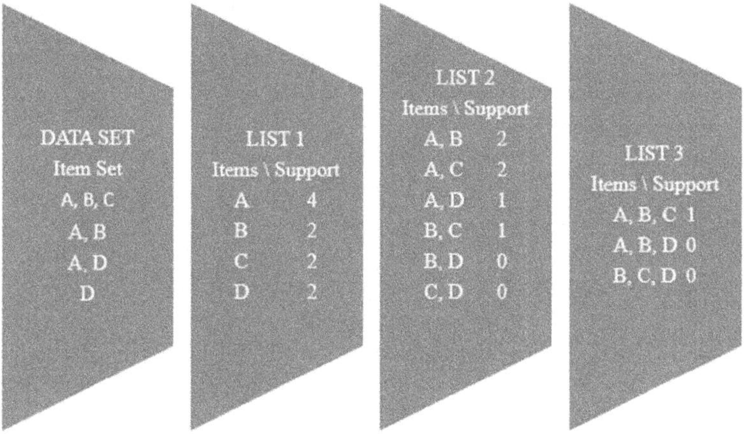

FIGURE 2.14    Data set and list 1, 2, and 3.

the impact of buying other products. It creates the probability of food items that we get, including the basket and list. In this way, it creates a template for such products, and this template is often a template call. Frequently used sample-based analysis techniques are implemented to find the probability of occurrence based on patterns.

Let's consider the example of a shopping site. Suppose we have items as A, B, C, and D. Based on customer need, item sets are prepared and based on customers who bought things together, LIST1, 2, and 3 are prepared using data mining algorithm, that is, apriori algorithm. Consider Figure 2.14 for data set and LIST 1, 2, and 3.

As mentioned earlier, the minimum support cost cannot be less than the minimum frequency of individual elements that can be seen in LIST 1. According to LIST 1, the minimum service frequency is "2". This means that at least the number of actions must be "2" or more than "2". In this example, we can set the Minimum Support Level to "2". Articles in LIST 2 are created by connecting each item located in LIST 1, and replicas being removed.

For LIST 2, support values are calculated from the whole data set, let's say, on checking the data set, A and B stand together 2 times (at the first and second row). Whilst B and D are never together, hence, support value for "B, D" is "0". According to the Minimum Support Value "A, D", "B, C", "B, D" and "C, D" has been removed from LIST 2. At LIST 3, the combination of LIST 2 and LIST 1 has found something like 3-way combination, and reputations are removed (Note: "A, B, C" and "B, A, C" gives same support value and that's why they need to be removed). Because every item set's support value in LIST 3 is less than Minimum Support, LIST 3 cannot be taken into consideration and, as result, final data points are found as LIST 2.

After all level's tables created confidence, values can be calculated using the following formula;

$$A \rightarrow B - \text{support } (A, B)/\text{support}(A) = (2/4) * 100 = 50\%$$

$$A \rightarrow C - \text{support } (A, C)/\text{support}(A) = (2/2) * 100 = 100\%$$

Basically, the above result shows the probability of existence of B (if A exists) is 50%. For "A ->"C", confidence value was found as 100%. On examining the dataset, it is clearly evident that every item set that includes "A" value, includes "C" value too. That's why the probability of the existence of "C" in presence of "A" in item-set is 100%. So, this is a basic example but, if you are working on huge transaction data of any e-commerce website, it becomes impossible to calculate these results manually. By considering the apriori algorithm scan dataset over and over again, the system/platform that apriori runs on should be fast enough. Due to these factors, Python, R, Spark are the proper environments to run apriori.

### 2.4.2.3 Principal Component Analysis (PCA)
The main idea behind this algorithm is to help reduce the size of the data set, which includes many variables either in a greater or lesser way, retaining all the variations up to the maximum extent. Repeating and transforming the variables to a whole new set is referred to as principal components. Principal components are orthogonal and the variable retention is decreased as we move down using the algorithmic properties. The first principal component solved using the algorithm retains maximum variance as it was in the original dataset. We can term principal components as eigen-vectors of a covariance matrix. Therefore, it shows orthogonal property.

The main goal of the PCA algorithm is to summarize the data on a relative scale. The results obtained using the algorithm are sensitive based on scaling. Steps for using PCA are as follows:

1. Standardization
   Mathematically, it can be represented as follows:

$$Z = \frac{Value - Mean}{Standard\ Deviation} \tag{2.5}$$

2. Compute the covariance matrix

$$|Cov(x,x)Cov(x,y)Cov(x,z)Cov(y,x)Cov(y,y)Cov(y,z)$$
$$Cov(z,x)Cov(z,y)Cov(z,z)| \tag{2.6}$$

3. In order to find out the principal components, calculate eigenvalues and eigenvectors

4. Calculate Feature Vector

   After discarding eigenvalues of lesser significance, the remaining principal components are arranged in a matrix referred to as Feature Vector.

5. Recast the Data Along the Principal Components Axes

   Resultant Data Set = Feature Vector * Standardized Original Data Set

### 2.4.2.4 Singular Value Decomposition

In linear algebra, one can make a call using singular value decomposition (SVD), since the distribution can be a real or complex matrix. This is a generalization of the eigenvalue, and it is the root of a positive semi definite normal matrix. This is done so that they look for a peak (e.g., consider a symmetric matrix that is not positive eigenvalues) to any m × n matrix via an extension that is just below the polar decomposition. It has many useful signal processing applications that should be taken into account in statistics.
   Single decay can be easily calculated using the following visual cues:

The left ventricles of the M are considered to be a group of orthonormal MM * eigenvectors.
The right ventricles of the M are actually an orthonormal set of MM eigenvectors.
The zero value of the non-zero M (found in the horizontal input of Σ) is considered to be the square root of the non-zero eigen values of both M M and MM *.

Apps that help to use SVD include pseudo inverse computation, small squares that include data, inseparable control, matrix measurement, and determine the distance, distance, and idle matrix.

### 2.4.2.5 Independent Component Analysis

Independent component analysis (ICA) is basically related to key analysis and asset analysis. ICA is considered to be the most powerful method.

The information analyzed by the ICA may be obtained from a variety of application platforms, which can include digital photographs, data, economic indicators and psychometric measurements. In most cases, these estimates are given to be considered as a set of similar symbols or a series of timelines. Typical examples are actually a mixture of speech signals taken from multiple microphones; these are brain waves that are recorded by multiple sensors and disturbing radio signals reach the cell phone, or perhaps a series of simultaneous signals obtained through some industrial processes.

## 2.4.3 Reinforcement Learning

The model in reinforcement learning (RL) represents the simulated dynamics of the environment from an agent's perspective. It provides a mapping of the state-action pair to the probability distribution of various states of the environment. Since it is not necessary for every RL agent to use a model, the most generalized categorization of RL algorithms is based on the algorithm being model-based or model-free. Various algorithms used for RL are grouped under one of these two categories, as shown in Figures 12.15–17.

Mathematically, the model can be represented as: $p(s_{t+1}|s_t, a_t)$ and the policy is $\pi_\lambda(a_t|s_t)$. System dynamics can be represented as:

$$p_\theta(\tau) = p_\theta(s_1, a_1, \ldots, s_N, a_N) = p(s_1)\prod_{t=1}^{N}\pi_\theta(s_t)p(s_t, a_t) \tag{2.7}$$

$$\theta maxE_{\tau \sim p_\theta(\tau)}\left[\sum_t r(s_t, a_t)\right] \tag{2.8}$$

Where $r(s_t, a_t)$ is the reward for action $a_t$.

Model based algorithms follow either one of the two approaches – Learn the Model and Given the Model or can be a hybrid of these approaches.

### 2.4.3.1 Learn the Model

In these model-based RL algorithms, a base policy $\pi(a_t|s_t)$ is run to make the model learn and the trajectory $\{(s_t, a_t, s_{t+1})_i\}$ is observed. Learning of a dynamics model $p(s, a)$ is performed with the objective of minimizing

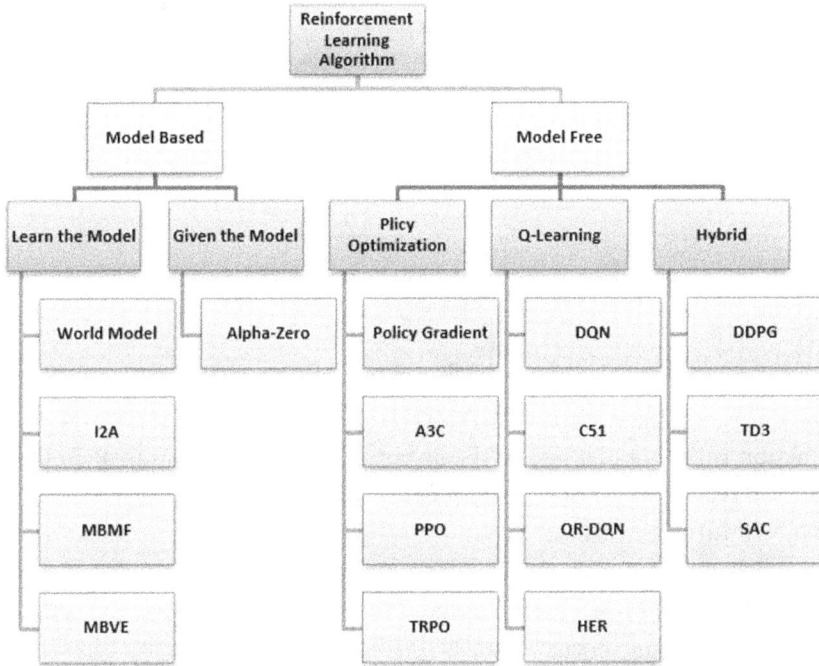

FIGURE 2.15   Classification of reinforcement learning algorithms.

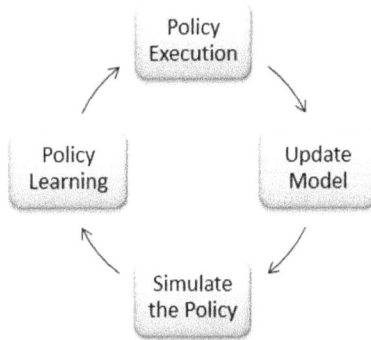

FIGURE 2.16   The process flow of reinforcement algorithm.

the overall value $\sum_i \left[ p(s_t a_t)_i - (s_{t+1})_i \right]^2$ Based on the dynamics, actions are selected.

The model is trained to minimize the squared error for the control function using sample learning data similar to supervised learning. The

FIGURE 2.17   The detailed process flow of reinforcement algorithm.

function value is a measure of the distance from target location and efforts spent to reach this point. Commonly used model-based RL approaches are explored further below.

2.4.3.1.1 World Model   The human brain prepares an internal model of the world around us based on the information perceived by the senses. This model, along with knowledge about the past, is used to take actions for the current state and also make predictions about the future. Similarly, in RL, the agent develops a world model and utilizes the representation of past and present knowledge for predictive modeling. The observations from the environment are processed by the World Model in two stages. The Vision Model encodes these observations from a high dimensional data into a low dimensional latent vector. This information is processed with the historic codes present in the memory stage, which is implemented using some form of neural network, to create a predictive representation capable of predicting future states. Representations from the Vision Model and Memory are compared by the controller to select the optimal or most suitable action.

2.4.3.1.2 Imagination Augmented Agent (I2A)   Successful implementation of model-based agents is limited to the domains where exact state transition models are available or to the low-dimensional environments which are easy to learn for the agent. In complex domains, the function approximation error is very high when a solution is attempted by using model-based agents with standard planning methods. I2A agents have been shown to overcome these shortcomings of model-based RL agents by approximate models which learn to interpret the imperfect predictions of such complex environments. This approach integrates model-free and model-based methods and thus outperforms model-free methods by using imperfect models.

2.4.3.1.3 Model-Based Priors for Model-Free Reinforcement Learning   This approach of model-based RL bridges the gap between model-based and model-free methods. It is based on a probabilistic framework which uses the cost estimated by model-based components as prior information for the model-free components. The model has been shown to overcome the bias and inaccuracy present in the dynamics model of the environment while retaining the fast convergence provided by model-based agents.

2.4.3.1.4 Model-Based Value Expansion   This is another hybrid RL algorithm, which attempts to reduce the sample complexity while the support for complex nonlinear dynamics is retained. The short-term horizon is simulated by using a dynamics model (model-based component), whereas long term values beyond the simulated horizon are estimated by using Q-learning (model-free component).

### 2.4.3.2 Given the Model – Alpha Zero Approach
The ability to learn themselves from the first principles is the most desired target for any AI agent. This superhuman capability was demonstrated by the AlpaGo Zero algorithm in the game of Go by defeating the best Go player in the world in October 2015. A more generalized version of this algorithm – AlphaZero – was proposed in 2018, which was tested on the games of Chess and Shogi.

### 2.4.3.3 Model-Free Reinforcement Learning
Unlike model-based RL, model-free RL algorithms do not use the transition model represented by transition probability distributions and the reward function associated with Markov decision process (MDP). Since "model" is a term used collectively for transition model and reward function, such algorithms are called model-free. These algorithms are based on a trial-and-error approach to optimize the policy, which can be grouped into two approaches: policy-based and value-based.

A policy has two types of value functions: (i) State value function, also known as V-value, and (ii) state-action value function, also known as Q-value. Different policies are compared based on their value functions to find the optimal or best policy. A policy "X" having value function "$V_x$" is considered to be better than the policy "Y" having value function "$V_y$" if $V_x > V_y$. Thus, the optimal policy gives two optimal value functions V-value$_{opt}$ and Q-value$_{opt}$. This is equivalent to finding V-value$_{opt}$ and Q-value$_{opt}$, which gives us the optimal policy. Thus, model-free RL algorithms can be implemented by two approaches:

(i) Direct or policy-based, also known as policy optimization or iteration – optimal value need not be calculated. Optimal policy obtained directly by the algorithm as shown in Figure 2.18.

(ii) Indirect or value-based, also known as Q-learning or value iteration – optimal state-action value is obtained, which is used to derive the optimal policy as shown in Figure 2.19.

### 2.4.3.4 Policy Optimization Approach
In these RL methods, the policy function maps the state directly to action, and the agent learns this policy function.

2.4.3.4.1 Policy Gradient Method   In this model-free RL method, the policy "$\pi$" having parameter "$\theta$" is trained to take actions "a" based on the observations. This policy can be deterministic, giving the specific action to be taken, or stochastic, giving probable chances of an action than can be taken.

The probability distribution of actions "a" for state space "s" can be written as

$$\Pi_\theta(a|s) = P\,[a|s]$$

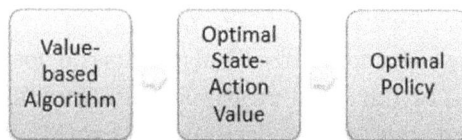

FIGURE 2.18   Direct RL model process.

FIGURE 2.19   Indirect RL model process.

For a discount factor "$\gamma$" and reward "r", z score optimization function $J(\theta)$ is maximized by obtaining the best value of the parameter "$\theta$".

$$J(\theta) = E_{\pi\theta}[\Sigma\gamma r]$$

The expected reward for a series of "N" steps is obtained by the summation of product of probability of trajectory "$\tau$" with its corresponding reward function.

$$J(\theta) = E\left[\sum_{t=1}^{N} R(s_t, a_t); \pi_\theta\right] = \sum_{\tau} P(\tau, \theta) r(\tau) \tag{2.9}$$

Where $\tau = (s_1, a_1, s_2, a_2, ..., s_N, a_N)$

For N $\square\infty$, the process runs indefinitely until it reaches the desired end state.

The objective of the policy gradient method is to obtain a policy "$\theta$" which creates a trajectory "$\tau$" such that the expected reward is maximized.

$$\theta max J(\theta) = \theta max \sum_{\tau} r(s_t, a_t) \tag{2.10}$$

Optimizing for maximum reward, the optimum value of parameter "$\theta$" written as $\theta_{opt}$ can be formulated mathematically as:

$$\theta max E_{\tau \sim P_\theta(\tau)}\left[\sum_{\tau} r(s_t, a_t)\right] \tag{2.11}$$

$$\text{Where } E_{\tau \sim P_\theta(\tau)}\left[\sum_{\tau} r(s_t, a_t)\right] = J(\theta) \tag{2.12}$$

For a continuous space, the expected value of a function f(x) can be written as

$$E_{x \sim P_\theta(x)} = \int P_\theta(x) f(x) dx \tag{2.13}$$

Also, the partial derivative $\nabla_\theta$ of a function f(x) can be written as:

$$\nabla_\theta f(x) = f(x)\frac{\nabla_\theta f(x)}{f(x)} = f(x)\nabla_\theta\left(logf(x)\right) \qquad (2.14)$$

The policy gradient $\nabla_\theta J(\theta)$ can, thus, be written as:

$$\nabla_\theta J(\theta) = \int\nabla_\theta\pi_\theta(\tau)r(\tau)d\tau = \int\pi_\theta(\tau)\nabla_\theta loglog\,\pi_\theta(\tau)r(\tau)d\tau \quad (2.15)$$

$$= E_{\tau\sim\pi_\theta(\tau)}\left[\nabla_\theta loglog\,\pi_\theta(\tau)r(\tau)\right] \qquad (2.16)$$

This policy gradient is used to update the policy.

2.4.3.4.2 Asynchronous Advantage Actor-Critic (A3C)   The A3C algorithm was introduced by Google's Deep Mind group in the year 2016 and it made the Deep Q Neural network (DQN) method obsolete. This algorithm is capable of working in both continuous and discrete environments. The 3 "A"s in the name describe the mechanism of this algorithm.

Asynchronous: In contrast to DQN, which is a single agent system implemented by single neural network, in A3C, multiple agents represented by multiple neural networks interact with the environment. This makes the learning more efficient. A3C operates in a global network and the multiple agents interact with their own copy of the environment through their own set of network parameters. Each agent experiences the environment independent of each other at the same time and the overall experience becomes more diverse compared to single agent DQN.

Actor-Critic: A3C integrates the benefits of value iteration and policy iteration methods by using Actor-Critic. It estimates both the value function and policy for a state "s" deployed as separate layers, which are fully connected and placed at the top of the network. The value estimate is used by the agent to update the policy, making the system more intelligent compared to the policy gradient method.

Advantage: In the policy gradient approach, the policy update rule informs the agents about its action being "good" or "bad". The network is updated accordingly such that a particular action is encouraged or discouraged based upon the discounted rewards or returns of past experiences. In the A3C approach, the advantage "A" is used rather than the discounted reward to update the policy. Using advantage, the agent can not only determine how good the actions were but also compare the reward with the expected value.

**2.4.3.4.3 Trust Region Policy Optimization (TRPO)**  A rule-based algorithm that can be used in environments with discrete or a continuous range. The TRPO updates the policy, to take all the best from the step-in order to improve the overall performance, while meeting the restrictions on how close to the old ones and the new ones' updated policy

**2.4.3.4.4 Proximal Policy Optimization (PPO)**  A target algorithm similar to a TRPO, which can perform in different or continuous areas of action. The PPO shares the encouragement with the TRPO in the work of answering this question: how can you accelerate policy development without the risk of a collapse? The idea is that the PPO enhances the stability of the actor training by limiting policy renewal at each training step.

PPO became very popular when open artificial intelligence made progress in Deep RL when they released an algorithm trained to play Dota2 and won against some of the top players in the world.

## 2.5  SUMMARY

Both the data and the effectiveness of the learning algorithms are essential for a successful ML model. Before the system can aid with intelligent decision-making, the sophisticated learning algorithms must be taught using real-world data and knowledge relevant to the target application. Because of its ability to learn from the past and make intelligent decisions, ML is gaining popularity in a variety of fields.

## REFERENCES

[1]  Schmidt, J., Marques, M. R., Botti, S., & Marques, M. A. (2019). Recent advances and applications of machine learning in solid-state materials science. *npj Computational Materials*, 5(1), 1–36.

[2]  Bojarski, M., Del Testa, D., Dworakowski, D., Firner, B., Flepp, B., Goyal, P., ... & Zieba, K. (2016). End to end learning for self-driving cars. arXiv preprint arXiv:1604.07316.

[3]  He, K., Zhang, X., Ren, S., & Sun, J. (2015). Delving deep into rectifiers: Surpassing human-level performance on imagenet classification. In Proceedings of the IEEE International Conference on Computer Vision(pp. 1026–1034).

[4] Liu, S. S., & Tian, Y. T. (2010, June). Facial expression recognition method based on Gabor wavelet features and fractional power polynomial kernel PCA. In International Symposium on Neural Networks (pp. 144–151). Springer.

[5] Waibel, A., & Lee, K. F. (Eds.). (1990). *Readings in Speech Recognition.* Elsevier.

[6] Pazzani, M., & Billsus, D. (1997). Learning and revising user profiles: The identification of interesting web sites. *Machine Learning,* 27(3), 313–331.

[7] Chan, P., & Stolfo, S. J. (1999). Toward scalable learning with non-uniform distributions: Effects and a multi-classifier approach. In Proceedings of the Fourth International Conference on Knowledge Discovery and Data Mining, 165–168.

[8] Guzella, T. S., & Caminhas, W. M. (2009). A review of machine learning approaches to spam filtering. *Expert Systems with Applications,* 36(7), 10206–10222.

[9] Huang, C. L., Chen, M. C., & Wang, C. J. (2007). Credit scoring with a data mining approach based on support vector machines. *Expert Systems with Applications,* 33(4), 847–856.

[10] Baldi, P., Brunak, S., & Bach, F. (2001). Bioinformatics: The Machine Learning Approach. MIT Press.

[11] Noordik, J. H. (Ed.). (2004). *Cheminformatics Developments: History, Reviews and Current Research.* IOS Press.

[12] Rajan, K. (2005). Materials informatics. *Materials Today,* 8(10), 38–45.

[13] Martin, R. M. (2020). *Electronic Structure: Basic Theory and Practical Methods.* Cambridge University Press.

[14] Hohenberg, P., & Kohn, W. (1964). Inhomogeneous electron gas. *Physical review,* 136(3B), B864.

[15] Oganov, A. R. (Ed.). (2011). Modern Methods of Crystal Structure Prediction. John Wiley.

[16] Walsh, A. (2015). The quest for new functionality. *Nature Chemistry,* 7(4), 274–275.

[17] Faber, F. A., Lindmaa, A., Von Lilienfeld, O. A., & Armiento, R. (2016). Machine learning energies of 2 million elpasolite (A B C 2 D 6) crystals. *Physical Review Letters,* 117(13), 135502.

[18] Wu, Y. J., Sasaki, M., Goto, M., Fang, L., & Xu, Y. (2018). Electrically conductive thermally insulating Bi–Si nanocomposites by interface design for thermal management. *ACS Applied Nano Materials,* 1(7), 3355–3363.

[19] Jalem, R., Kanamori, K., Takeuchi, I., Nakayama, M., Yamasaki, H., & Saito, T. (2018). Bayesian-driven first-principles calculations for accelerating exploration of fast ion conductors for rechargeable battery application. *Scientific Reports,* 8(1), 1–10.

[20] Rosenblatt, F. (1958). The perceptron: A probabilistic model for information storage and organization in the brain. *Psychological Review*, 65(6), 386.

[21] McCulloch, W. S., & Pitts, W. (1943). A logical calculus of the ideas immanent in nervous activity. *The Bulletin of Mathematical Biophysics*, 5(4), 115–133.

[22] Ye, W., Chen, C., Wang, Z., Chu, I. H., & Ong, S. P. (2018). Deep neural networks for accurate predictions of crystal stability. *Nature Communications*, 9(1), 1–6.

[23] Ren, Z., & Lee, Y. J. (2018). Cross-domain self-supervised multi-task feature learning using synthetic imagery. In Proceedings of the IEEE Conference on Computer Vision and Pattern Recognition (pp. 762–771).

[24] Seko, A., Hayashi, H., Nakayama, K., Takahashi, A., & Tanaka, I. (2017). Representation of compounds for machine-learning prediction of physical properties. *Physical Review B*, 95(14), 144110.

## Web References

1. https://machinelearningmastery.com/basic-concepts-in-machine-learning/
2. https://medium.com/technology-nineleaps/popular-machine-learning-algorithms-a574e3835ebb
3. www.newtechdojo.com/list-machine-learning-algorithms/
4. www.analyticsvidhya.com/blog/2017/09/common-machine-learning-algorithms/
5. www.intellspot.com/decision-tree-examples/
6. https://software.intel.com/content/www/us/en/develop/articles/mathematical-concepts-and-principles-of-naive-bayes.html
7. https://dataaspirant.com/naive-bayes-classifier-machine-learning/
8. www.analyticsvidhya.com/blog/2021/03/beginners-guide-to-support-vector-machine-svm/
9. www.javatpoint.com/machine-learning-random-forest-algorithm
10. https://medium.com/capital-one-tech/random-forest-algorithm-for-machine-learning-c4b2c8cc9feb
11. https://in.springboard.com/blog/what-is-linear-regression/
12. https://medium.com/analytics-vidhya/ordinary-least-square-ols-method-for-linear-regression-ef8ca10aadfc
13. www.analyticsvidhya.com/blog/2019/08/comprehensive-guide-k-means-clustering/
14. https://dataaspirant.com/k-means-clustering-algorithm/
15. https://sergencansiz.medium.com/what-is-apriori-algorithm-and-how-it-works-c99a9513753d

16. https://builtin.com/data-science/step-step-explanation-principal-compon ent-analysis
17. https://smartlabai.medium.com/reinforcement-learning-algorithms-an-intuitive-overview-904e2dff5bbc
18. https://jonathan-hui.medium.com/rl-policy-gradients-explained-9b13b 688b146
19. https://medium.com/emergent-future/simple-reinforcement-learning-with-tensorflow-part-8-asynchronous-actor-critic-agents-a3c-c88f7 2a5e9f2
20. https://smartlabai.medium.com/reinforcement-learning-algorithms-an-intuitive-overview-904e2dff5bbc
21. https://towardsdatascience.com/reinforcement-learning-explained-visua lly-part-3-model-free-solutions-step-by-step-c4bbb2b72dcf

# CHAPTER 3

# Applications of Machine Learning and Deep Learning

Gagandeep Kaur,[1] Satish Saini[1]
and Amit Sehgal[2]

[1]RIMT University, Punjab, India
[2]Sharda University, Uttar Pradesh, India
E-mail: kaurgagan10deep@gmail.com; satishsainiece@gmail.com
amitsehgal26@gmail.com

## 3.1 MACHINE LEARNING APPLICATIONS

Machine learning has become a short-hand for modern technologies, and it is growing briskly. Without even knowing, we are using machine learning in our daily lives, for example: Alexa, Google Assistant, Google Maps and so on. Figure 3.1 depicts the popular day-to-day applications of machine learning.

### 3.1.1 Image Recognition

This is the most widely used application of machine learning. It is used for the identification of digital images, objects, places, and so on. Popular features include face and image recognition, and automated friend suggestions. For example, Facebook offers an automatic feature (the suggestion of tagging friends). If we upload photos of our friends to Facebook, we will have a tagging system, a suggestion, a name, and a technology based on facial

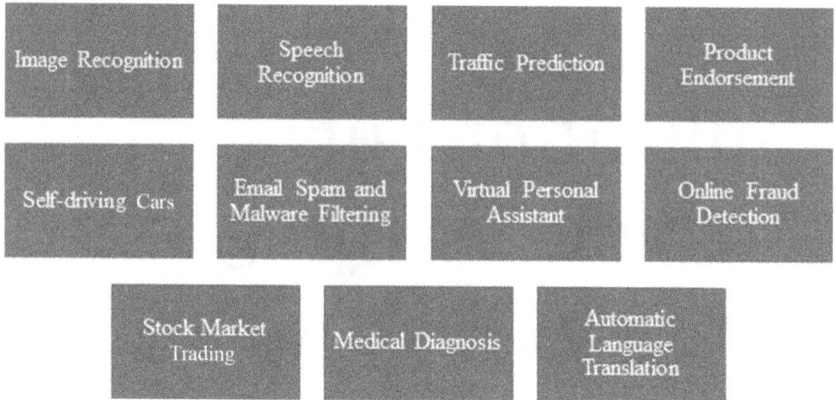

| Image Recognition | Speech Recognition | Traffic Prediction | Product Endorsement |
| Self-driving Cars | Email Spam and Malware Filtering | Virtual Personal Assistant | Online Fraud Detection |
| Stock Market Trading | Medical Diagnosis | Automatic Language Translation |

FIGURE 3.1   Machine learning applications.

recognition and machine learning algorithms. It is all based on Facebook's "Deep Face" project.

### 3.1.2  Speech Recognition

Google has the ability to find your voice, "voice recognition", and it's become a popular machine learning app. Speech recognition is the process of converting voice messages to text, and is also known as "speech-to-text conversion". ML algorithms have been extensively used in speech-recognition systems since they were developed. Siri, Google Assistant, Alexa – these all-use voice or speech recognition technology in order to follow instructions.

### 3.1.3  Traffic Prediction

Suppose we want to move to some unfamiliar location; we will use the help of the Google Maps app. It will show us the correct and shortest route with traffic prediction. In the case of traffic moving slowly or when there is heavy congestion, it will use the following two methods:

a. Real-time vehicle location in the form of the Google Map app and sensors.

b. The average time it took to do this in recent days and at the same time.

Whosoever is using this application can help it to perform better because it accepts user information and feeds it to the database in order to improve the overall performance.

### 3.1.4 Product Endorsement

Numerous entertainment and e-commerce companies, such as Netflix and Amazon, use ML algorithms by recommending the product to the user. Due to ML algorithms, we start seeing ads of the same product which we searched on Google, Amazon and other e-commerce apps.

### 3.1.5 Self-driving Cars

Self-driving cars is considered to be the most exciting application of ML. It is crucial in the development of self-driving vehicles. With the unsupervised learning method, you can teach cars and people to detect objects in motion.

### 3.1.6 Email Spam and Malware Filtering

After receiving any new email, it gets filtered out automatically. F or example, important, social or spam. Spam is the most important area using ML algorithms for detecting and filtering out malware.

### 3.1.7 Virtual Personal Assistant

As the name suggests, virtual personal assistants like Siri, Alexa, Google assistant, and so on, help us to find information using voice instructions. These devices respond in a variety of ways, according to voice commands, such as playing music, making phone calls, opening or sending emails to make an appointment, and so on. Virtual screens, the application of machine learning algorithms, are an important part of this.

### 3.1.8 Online Fraud Detection

ML provides an advantage for online transactions, which are safe and secure thanks to fraud detection in the transaction. When we run a series of online transactions, there can be several different ways to make fraudulent transactions, such as fake invoices, fake documents, and money-stealing in the middle of the transaction. So, to check this, we need a neural feedback network that will help us make sure that this is the real deal or to spot any unfair transactions. For any legitimate trade, the output data will be included in some hash values, and these values are the input data for the next round. For such transactions, there is a certain pattern, which changes for fraudulent transactions, making online transactions are more secure, and providing reassurance.

### 3.1.9 Stock Market Trading

ML is widely used in exchange trading. On the stock exchange, there is always risk, and ups and downs in promotions. Machine learning uses a neural network with long short-term memory to predict trends in the stock market.

### 3.1.10 Medical Diagnosis

In the field of medicine, machine learning can be used to diagnose diseases. Due to the fact that medical technology is developing very quickly and is able to build a 3D model that can predict the exact position due to brain damage, it helps in detecting brain tumors and other brain-related diseases, and is more convenient than invasive procedures.

### 3.1.11 Automatic Language Translation

Now, if we visit unfamiliar places without knowing the language of that specific region, it won't be a problem because machine learning will help us through the process of converting the unknown text to any of our familiar languages. Google Neural Machine Translation provides this feature, which is Instagram machine learning that translates text into the desired language, and is also known as automatic translation. This is technology that uses a sequence-to-sequence learning algorithm together with the function of recognizing and translating text from one language to another.

## 3.2 DEEP LEARNING

Deep learning is a branch of machine learning. It is inspired by the functions and structure of the brain, so-called artificial neural networks [1]. Following this, the deep learning algorithms are trying to get to the same conclusion as the person who received it, and are continually analyzing data with a specific structure. In order to achieve this goal, deep learning is aided by a hierarchical structure [1, 2] of algorithms called neural networks.

The shape of a neural network is based on the shape of the human brain. In the same way that we use our brains to identify and classify distinct data or information, a neural network can be trained to carry out identical tasks with statistics. Each layer of neural networks may be considered as a filter because it works to provide superb results. The human mind works in the same way. When we get hold of new data, the brain tries to examine this with known things; the concept of deep learning is the same.

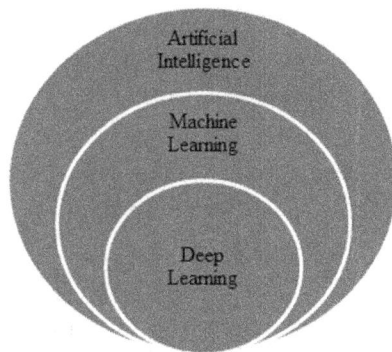

FIGURE 3.2    Relation between AI, ML and DL.

Neural networks allow you to carry out various functions, such as regression, grouping or classification. The data in neural networks is based on the similarity index between the samples taken. Neural networks, in general, may perform the same functions as ML algorithms but this is not entirely true in every case.

Deep learning models with artificial neural network support are able to solve complex problems [3, 4] which can never be solved with the use of ML algorithms. This is possible due to the latest technological advancements in artificial intelligence. Chatting through chatbots, self-driving and talking to Alexa is only possible due to deep learning [5–8].

Deep learning is responsible for today's industrial revolution. Therefore, it is the best and easiest approach for building real-time machine intelligence that we've had so far.

## 3.3 MACHINE LEARNING VS. DEEP LEARNING

Deep learning being a subset or branch of machine learning, then, what's the actual difference between them? The actual difference is based on the nature of data being used and the learning methods.

ML algorithms use structured data and estimate results. We can also say that on the basis of input data, specified features are determined. Usually, a series of pre-processing should be arranged in an organized manner.

Deep learning, which excludes some preprocessing of data, is typically used in machine learning. These algorithms are capable of intercepting and processing unstructured data [9–11], such as text and images, as well as automating the feature extraction process, eliminating some dependence on human experts. For example, suppose we have several images of different animals, and we want to classify them just into "cat", "dog", "hamster" and so on. Deep learning algorithms can figure out which of the traits (such as ears!) are the most significant, and which should be discriminated from one animal to the next. This is a hierarchy of functions in machine learning that is manually set by a human expert. Then there's gradient descent and back propagation, as well as a deep learning algorithm that adapts to the environment.

Different types of learning, such as supervised learning, unsupervised learning and reinforcement learning can be used by machine learning and deep learning models [3, 5, 8, 9]. Data is used for classification or prediction in supervised labelled learning, which necessitates human intervention to appropriately label the input data. Unsupervised science, on the other hand, does not require labelled datasets, and, instead, finds patterns in the data, groups them and assigns unique characteristics to them. Reinforcement learning is the process by which a model improves its accuracy in order to conduct actions in an environment where the goal is to maximize profits.

## 3.4 HOW DEEP LEARNING WORKS

Using a combination of input weights and biases, deep learning or artificial intelligence networks seek to replicate the human brain [12, 13]. These components work in tandem to effectively identify, classify, and describe database objects.

Deep neural networks are made up of numerous layers of interconnected nodes [10–13], each of which builds on the preceding layer and regulates prediction or classification optimization exactly. A network direct jump is used to do this type of calculation. The input and output layers of a deep neural network are the layers that are visible. In the input layer, the deep learning model generates data for processing, and in the output layer, the final classification prediction is created.

Back propagation is a technique that involves using algorithms like gradient descent to determine prediction errors and then adjusting the weights and biases of the features by travelling back through the layers, attempting to train the model. The use of both forward and back propagation allows for accurate correction of any flaws existing in the predicted neural network. After some time, the system becomes increasingly accurate.

## 3.5 APPLICATIONS OF DEEP LEARNING

Deep learning applications, being part of our everyday life, are usually so effectively integrated with products and services that their consumers are unaware of the problems of data processing running in the background. The following are a few examples of applications.

### 3.5.1 Law Enforcement

Deep learning algorithms are able to assess and draw conclusions from transaction data in order to spot potentially fraudulent or criminal tendencies. Vast amounts of data must be swiftly and fully evaluated in order to maximize the efficiency and efficacy of research and analysis of data patterns, as well as evidence, audio and video recordings, photos, images and legal documents.

### 3.5.2 Financial Services

Predictive analytics is frequently used by financial institutions to handle algorithmic stock trading, assess credit selection risks, detect fraud, and assist in the management of clients' portfolios.

### 3.5.3 Customer Service

In their customer support procedures, many companies deploy deep learning technologies. Chatbots are the most common type of AI and can be found in a variety of applications, websites and social media platforms. Traditional chatbots employ natural language, and many call centre menus include a visual signal. More complex chatbot solutions, on the other hand, try to figure out if there are several answers to shotgun questions. Based on the responses it receives, the chatbot tries to answer these queries directly or transfers the connection to a human user. The most well-known examples are Google Assistant and Alexa. This expands the concept of a chatbot to incorporate voice recognition, ushering in a new era of client engagement.

### 3.5.4 Healthcare

The medical field has made use of the deep learning capabilities of digitizing data, records and photos. Medical imaging professionals, medical practitioners, image recognition tools and services help them to study and evaluate more images in less time.

| Convolutional Neural Networks (CNNs) | Long Short Term Memory Networks (LSTMs) | Recurrent Neural Networks (RNNs) | Generative Adversarial Networks (GANs) |
|---|---|---|---|
| Radial Basis Function Networks (RBFNs) | Multilayer Perceptrons (MLPs) | Self Organizing Maps (SOMs) | Deep Belief Networks (DBNs) |
| Restricted Boltzmann Machines (RBMs) | Autoencoders | | |

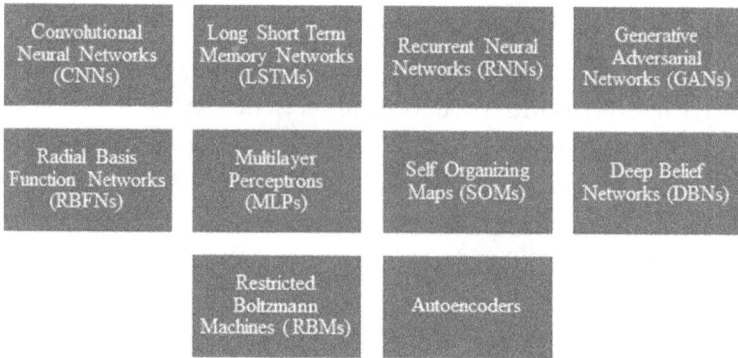

FIGURE 3.3    List of deep learning algorithms.

## 3.6  DEEP LEARNING ALGORITHMS

The ten most popular algorithms used in deep learning are outlined in Figure 3.3.

Deep learning algorithms can be applied to any data type, which necessitates a significant amount of processing power and data to tackle complicated issues. The algorithms are explained one-by-one as follows:

### 3.6.1  Convolutional Neural Networks (CNNs)

CNNs, commonly referred to as ConvNets, are multilayer neural networks that are mostly used for object detection and image processing. In 1988, Yann LeCun, then known as LeNet, became CNN's first correspondent. It's used to recognize characters like postal codes and digits, among other things. As expected, CNNs are widely employed for identifying satellite data, medical image processing, and series and anomaly detection as explained in Figure 3.4. CNNs contain numerous layers for processing and extracting features, which are as follows:

Convolution Layer: This layer is responsible for performing convolution operation through various filters.

Rectified Linear Unit (ReLU): This layer provides output in a rectified feature form by performing several operations on its elements.

Pooling Layer: The pooling layer receives the output of ReLU for further processing. Pooling is a technique for reducing the dimensionality of a feature map by down-selecting features. It turns the two-dimensional arrays that arise as a result of combined feature map's applications into a single, continuous, linear vector through optimization.

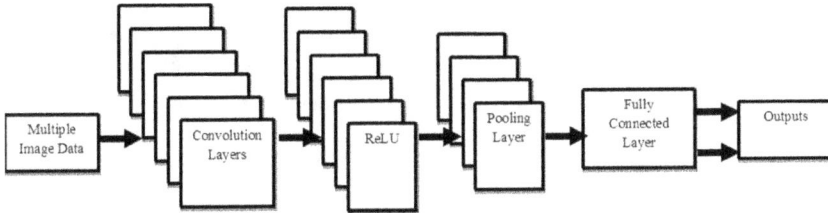

FIGURE 3.4    CNN algorithm layers.

Fully Connected Layer: To categorize and identify the images, a fully connected layer is established when a new matrix is generated from the pooling layer and acts as an input [13–14].

## 3.6.2 Long Short-Term Memory Networks (LSTMs)

An LSTM network is a sort of recurrent neural network (RNN) that is capable of learning and remembering long-term consistency. The default behaviour is to remember the past and information over a long period of time. LSTMs are data storage devices that keep data for a set amount of time. Because they recall their prior entries, they can be utilized in time series forecasting. The four interaction layers interact in a unique way in LSTM centres, which have a chain structure. In addition, LSTM is extensively used to anticipate the centre's time series in voice recognition, music composition and drug research.

To begin with, LSTMs ignore irrelevant elements of past states. Following that, they selectively update the cell-state values, followed by the output of specific cell-state elements [15–17].

## 3.6.3 Recurrent Neural Networks (RNNs)

The LSTM output is fed as input to the current step through elements arranged in the form of directed loops. The LSTM centre's output is the phase's entry and the ability to remember prior records thanks to its built-in memory. Time series analysis, image captioning, handwriting identification, machine translation and natural language processing are all common applications for RNNs.

In an RNN, the output at time t-1 is present at the input at time t. The output from time t is also fed into time t+1's input. Zones of any length can be handled by RNNs. The formula takes into consideration past data, and the project's input data should not grow in size. The RNN feature is demonstrated in Figure 3.5.

FIGURE 3.5   Working of Google's autocompleting feature.

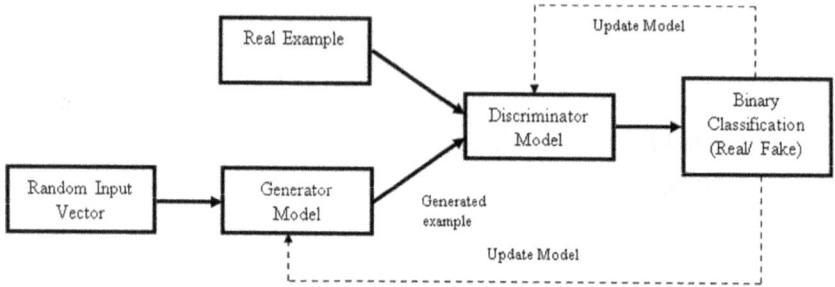

FIGURE 3.6   Working of GANs algorithm.

## 3.6.4 Generative Adversarial Networks (GANs)

GANs are those deep learning algorithms which are responsible for generating fresh copies of existing data, such as training data. A GAN is made up of two parts: a discriminator that learns about false data and a generator that learns how to make fake data. Everything's utilization rises over time. They can be used to improve astronomy photographs as well as imitate dark matter gravitational lensing. In computer games, designers use full size, low resolution 2D textures that are restored to 4K or higher quality by clicking on preview-training. All of them aid in the creation of realistic photos and movies, as well as animated images of the human face and the visualization of 3D objects.

The discriminator in this algorithm teaches how to tell the difference between genuine and misleading information, as well as real data samples. The generator generates phoney data during the initial session, and the receiver quickly learns to reject it. The technique then updates the model by sending the results to the generator and discriminator. The GANs method is depicted in Figure 3.6.

## 3.6.5 Radial Basis Function Networks (RBFNs)

Radial, primary and activation functions are used in RBFNs, which are a type of feedback neural network. An input layer, a hidden layer and an

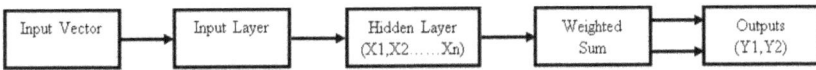

FIGURE 3.7    Example of an RBFN.

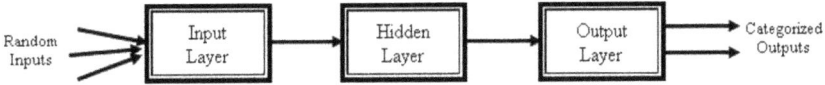

FIGURE 3.8    Example of MLP algorithm.

output layer comprise the three layers. Regression, classification and time series prediction are all performed using these layers. The similarity of the inputs in the models from the training set is used by RBFNs to distinguish them. The input layer of RBFNs is fed via an input vector which has a variety of RBF neurons as shown in Figure 3.7. The function gets a weighted amount of input, with one location per category or data category in the output layer. Gaussian transfer functions are present in the buried layer, and their effects are inversely proportional to the distance from the neuron's centre. The output of the network consists of a combination of linear radial-based input functions along with neuron parameters.

## 3.6.6 Multilayer Perceptrons (MLPs)

MLPs are a great place to jump off if you want to learn more about in-depth learning technology. Multiple layers of perceptrons with functional capabilities make up MLPs in the feed-forward neural network category. MLPs have a complete coupling between their input and output layers. There are the same number of input and output layers, but they can have several hidden layers, allowing them to be used to construct image recognition, machine translation software and speech recognition [17–19].

The network input layer is where MLPs ingest data. The signal goes to one side through layers of neurons connected to the graph. MLPs incorporate all inputs and weights that exist between both the layers. Activation functions are used by MLPs to decide which nodes should be active. ReLUs, sigmoid functions and tanh are all activation functions. MLPs teach the model to comprehend integration and to rely on independent and targeted variables from the training data set. Figure 3.8 shows an example of the MLP method in action, random input images after computing weights and implementing appropriate functions are categorized into specified outputs.

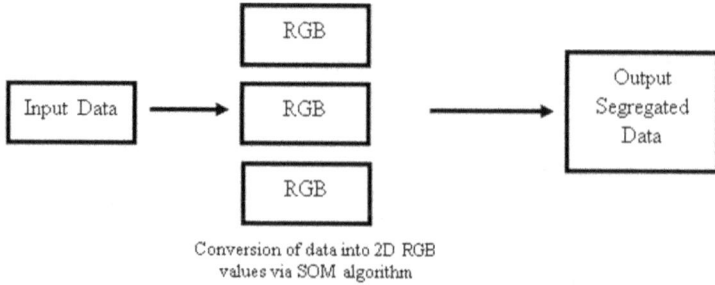

Conversion of data into 2D RGB
values via SOM algorithm

FIGURE 3.9    Example of SOMs.

## 3.6.7 Self-organizing Maps (SOMs)

SOMs were invented by Professor Teuvo Kohonen, and use artificial neural networks to minimize data size and allow data perception. The difficulty of individuals being able to see the most advanced data is addressed by data recognition. SOMs are intended to assist users in comprehending this complex information.

In SOMs, the weight of each node is determined and a vector is selected from the training data at random. Each node is examined to determine the most favorable weights to be the input vector. The successful node will be termed as The Best Matching Unit (BMU). Under the BMU, these algorithms gain traction, and the number of neighbours decreases. They also give the sample's vector the winning weight. When a node approaches a BMU, the change in its weight is even greater. The farther the neighbour gets away from the BMU, the more it learns. The second step in the production of N is repeated by SOMs. Figure 3.9 shows the input vector diagram which contains various colors. This information is fed into the SOM, in which the input is converted into 2D RGB values before separating the different colours.

## 3.6.8 Deep Belief Networks (DBNs)

DBNS models are generative models with multiple layers of stochastic, latent variables. Option values are hidden variables that are normally concealed from units. Each layer of a DBN contains an RBM that communicates with both the adult and subsequent levels, and each layer has a collection of Boltzmann machines with connections between two layers. Deep Belief Networks are capable of capturing images, videos and motion data.

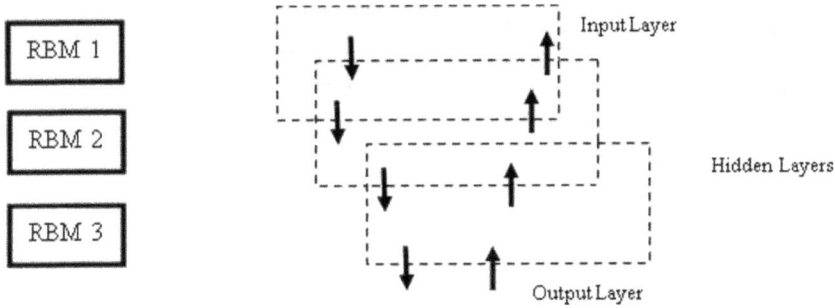

FIGURE 3.10    Example of DBN architecture.

DBNs are trained using greedy learning methods. To learn the generative top-down weights, the algorithm employs a layer-by-layer technique. Gibb's sampling is used to draw a sample from the concealed layers, that is, the two top layers. DBNs draw the sample using the visible units. The model employs a single pass for ancestral sampling. DBNs learn that one bottom-up pass can determine the values of tiny fluctuations in each layer. An example of DBN architecture is shown in Figure 3.10.

### 3.6.9 Restricted Boltzmann Machines (RBMs)

RBM is a Geoffrey Hinton-designed stochastic neural network which can learn the probability distribution for a set of input data. This approach is employed in a variety of applications, including dimensionality reduction, classification, regression, general filtering, symptom learning and subject mathematical modelling.

There are two layers to RBMs: visible and buried units as explained in Figure 3.11. All of the units, visible and concealed, are linked together. RBMs have no output nodes since the bias unit joins all of the visible and hidden units together. There are two phases to the RBM algorithm: forward and backward pass. In RBM algorithms, inputs are accepted and converted into a set of numbers, which it then encoded in the forward pass. Each input is combined with a weight as an independent factor and a single overall bias by the algorithm. After that, the output is passed to the hidden layer.

RBM converts the set of numbers into reconstructed inputs in the backward pass. Every activation is combined with weight as an individual factor as well as overall bias in the algorithm. Then the output is subsequently passed to the visible layer for reconstruction. In order to assess the outcome quality, the original signal and the reconstructed signal are compared here.

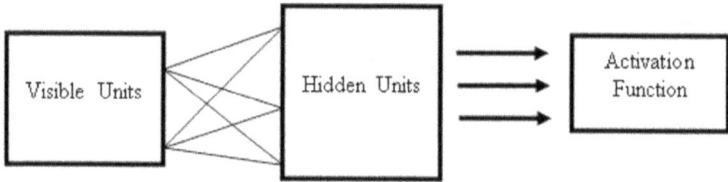

FIGURE 3.11   Working of RBM.

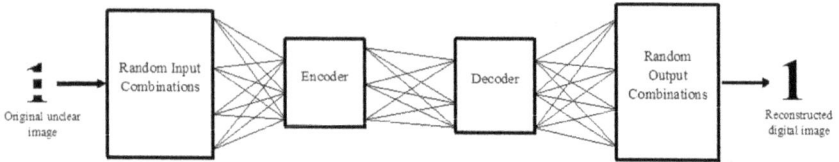

FIGURE 3.12   Operation of autoencoders.

### 3.6.10 Autoencoders

Autoencoders are a sort of neural network with feedback, which have identical input and output data. They were created by Geoffrey Hinton in 1980 with the goal of tackling unsupervised learning challenges. They were trained to copy all of the data to the output layer from the input layer using neural networks. Image processing, popularity prediction and pharmaceutical discovery are some of the other applications for autoencoders.

The encoder, decoder and the code are the three essential components of the autoencoder. Autoencoders are set up to look for input and convert it to a different format. After that, they try to recreate the original input as accurately as possible. The neural autoencoder network is used when the digital image is not clearly apparent. Autoencoders encode an image first, then lower the input size to a more manageable amount. Finally, by determining the image, the autoencoder reconstructs the image. The operation of autoencoders is depicted in Figure 3.12.

### 3.7 SUMMARY

Deep learning is a hot topic in machine learning; it is defined as a set of nonlinear layers that learn many levels of data representations. Researchers and experts in the field of machine learning have attempted to derive data and patterns representations from raw data. Unlike standard techniques of machine learning and data mining, high-level data representations are generated from massive volumes of raw data in the case of deep learning. As a result, deep learning has been used to solve a wide range of real-world issues.

## REFERENCES

[1] Shrestha, A., & Mahmood, A. (2019). Review of deep learning algorithms and architectures. *IEEE Access*, 7, 53040–53065.

[2] Pouyanfar, S., Sadiq, S., Yan, Y., Tian, H., Tao, Y., Reyes, M. P., ... & Iyengar, S. S. (2018). A survey on deep learning: Algorithms, techniques, and applications. *ACM Computing Surveys (CSUR)*, 51(5), 1–36.

[3] Najafabadi, M. M., Villanustre, F., Khoshgoftaar, T. M., Seliya, N., Wald, R., & Muharemagic, E. (2015). Deep learning applications and challenges in big data analytics. *Journal of Big Data*, 2(1), 1–21.

[4] Goodfellow, I., Bengio, Y., Courville, A., & Bengio, Y. (2016). *Deep Learning* (Vol. 1, No. 2). MIT Press.

[5] Wang, H., Chen, S., Xu, F., & Jin, Y. Q. (2015, July). Application of deep-learning algorithms to MSTAR data. In 2015 IEEE International Geoscience and Remote Sensing Symposium (IGARSS) (pp. 3743–3745). IEEE.

[6] Pedrycz, W., & Chen, S. M. (Eds.). (2020). *Deep Learning: Algorithms and Applications*. Springer.

[7] Abdel-Hamid, O., Mohamed, A. R., Jiang, H., Deng, L., Penn, G., &Yu, D. (2014). Convolutional neural networks for speech recognition. *IEEE/ACM Transactions on Audio, Speech, and Language Processing*, 22(10), 1533–1545.

[8] Rosenblatt, F. (1958). The perceptron: A probabilistic model for information storage and organization in the brain. *Psychological Review*, 65(6), 386.

[9] LeCun, Y., Bengio, Y., & Hinton, G. (2015). Deep learning. *Nature* (2015). May, 521 (7553): 436 10.1038/nature14539.

[10] Gheisari, M., Wang, G., & Bhuiyan, M. Z. A. (2017, July). A survey on deep learning in big data. In 2017 IEEE international conference on computational science and engineering (CSE) and IEEE international conference on embedded and ubiquitous computing (EUC) (Vol. 2, pp.173–180). IEEE.

[11] Pouyanfar, S., Sadiq, S., Yan, Y., Tian, H., Tao, Y., Reyes, M. P., ... & Iyengar, S. S. (2018). A survey on deep learning: Algorithms, techniques, and applications. *ACM Computing Surveys (CSUR)*, 51(5), 1–36.

[12] Vargas, R., Mosavi, A., & Ruiz, R. (2017). Deep learning: A review. *Advances in Intelligent Systems and Computing*, 5(2), 1–12.

[13] Hinton, G. E., Osindero, S., & Teh, Y. W. (2006). A fast learning algorithm for deep belief nets. *Neural Computation*, 18(7), 1527–1554.

[14] Zeiler, M. D., & Fergus, R. (2014, September). Visualizing and Understanding Convolutional Networks. In European Conference on Computer Vision (pp. 818–833). Springer.

[15] Ker, J., Wang, L., Rao, J., &Lim, T. (2017). Deep learning applications in medical image analysis. *IEEE Access*, 6, 9375–9389.

[16] Hatcher, W. G., &Yu, W. (2018). A survey of deep learning: platforms, applications and emerging research trends. IEEE Access, 6, 24411–24432.

[17] Ahmad, J., Farman, H., & Jan, Z. (2019). Deep learning methods and applications. In Deep Learning: Convergence to Big Data Analytics (pp. 31–42). Springer.

[18] Deng, L. (2014). A tutorial survey of architectures, algorithms, and applications for deep learning. *APSIPA Transactions on Signal and Information Processing*, 3.

[19] Manaswi, N. K., Manaswi, N. K., & John, S. (2018). *Deep Learning with Applications Using Python* (pp. 171–197). Apress.

## Web References

1. www.javatpoint.com/applications-of-machine-learning
2. www.edureka.co/blog/machine-learning-applications/
3. www.simplilearn.com/tutorials/machine-learning-tutorial/machine-learning-applications
4. www.mygreatlearning.com/blog/what-is-machine-learning/
5. www.simplilearn.com/tutorials/deep-learning-tutorial/deep-learning-algorithm
6. www.ibm.com/cloud/learn/deep-learning
7. https://towardsdatascience.com/what-is-deep-learning-and-how-does-it-work-2ce44bb692ac

# CHAPTER 4

# Environmental Monitoring in Wireless Sensor Networks using AI

Arpana Mishra[1] and Rashmi Priyadarshini[2]

[1]*Sharda University, Greater Noida, Uttar Paradesh, India,*
*shubhparna20@gmail.com)*

[2]*Sharda University, Greater Noida, Uttar Pradesh, India,*
*rashmi.priyadarshini@sharda.ac.in*

## 4.1 INTRODUCTION OF ENVIRONMENTAL MONITORING

Ecological monitoring is an apparatus to evaluate natural conditions and patterns, support strategy advancement and its execution, and foster data for answering to public policymakers, worldwide discussions, and general society.

Over the previous decade, few of nations of Europe and Central Asia have had the option of keeping up with existing checking exercises. The checking of metropolitan air contamination – a significant hazard to human wellbeing – is poor in numerous urban communities. Observing dangerous waste is ineffective and modern discharges are not necessarily routinely checked, reducing the adequacy of strategy instruments, for example, emissions charges and fines. The observation of transboundary air contamination likewise needs reinforcing. Also, numerous European and Central Asian nations need uniform public approaches across various checking regions; their existing order frameworks are frequently contrary to global principles.

At the Fifth Environment for Europe Conference (Kyiv, 2003), UNECE Ministers embraced suggestions for fortifying natural checking and data frameworks in European and Central Asian nations arranged by the UNECE Working Group on Environmental Monitoring. Ayodele additionally embraced the UNECE rules for the readiness of public condition of the climate reports. Together, these archives give a guide to the reinforcement of observing and reporting in the European and Central Asian sub-region [1].

## 4.2 APPLICATIONS OF WIRELESS SENSOR NETWORK (WSN)

### 4.2.1 Air Monitoring

Air contaminants are known for their unfavorable impact on human health and the environment. Emissions of nitrogen oxides and non-methane unpredictable natural mixtures are the primary driver of the development of ground-level ozone, which affects human health and the environment. The air poisons pointer surveys pressures from explicit contaminants on climatic air across singular nations, and also distinguishes pressures from specific public areas like energy, transport, modern cycles, farming and waste administration. Based on this pointer, public specialists can change the public ecological strategy by, for example, amending discharge principles and outflow limits Data on toxin discharges is vital for the appraisal of transboundary air contamination and for worldwide collaboration to resolve this issue [2].

### 4.2.2 Water Monitoring

Inexhaustible freshwater assets have major natural and monetary worth. Their circulation fluctuates generally among and inside nations. Pressure on freshwater assets is applied by overexploitation and by contamination. Water resources relating assets deliberation to restoration of stocks is a focal issue in maintainable freshwater across the board. Where a critical portion of a nation's water comes from transboundary streams, pressures between nations can emerge, particularly if water accessibility in the upstream nation is more prominent than in the downstream one. The Convention on the Protection and Use of Transboundary Watercourses and International

Lakes necessitates that the Parties present practical use of water, including a biological system approach and the levelheaded and reasonable utilization of transboundary waters [3].

### 4.2.3 Biodiversity

Economical advancement relies upon a sound climate, which, in turn, relies upon biological system variety. Secured regions, particularly within the full scope of the International Union for Conservation of Nature Protected Area Categories, are fundamental for rationing biodiversity [4].

The biodiversity marker provides a way to gauge the reaction to the debasement of environments and the deficiency of biodiversity in a country [5]. It exhibits the degree to which regions significant for preserving biodiversity, social legacy, logical examination, entertainment, normal asset support and other natural qualities are shielded from inconsistent employments [6].

### 4.2.4 Waste Monitoring

Waste addresses an extensive loss of assets as materials and energy. The treatment and removal of waste may cause ecological contamination and open people to destructive substances. A decrease in the volume of waste created is subsequently a sign of an economy's move towards less material-concentrated creation and utilization designs. The principle reason for the waste pointer is to quantify the pressure on the climate of the aggregate sum of produced waste and waste by class. The waste power addresses a main thrust pointer and shows reaction to anthropogenic exercises [7].

### 4.2.5 Distant Sensing

Distant detecting can play a significant part in checking and writing about natural issues, specifically when the objective of such perceptions is to survey toxin impacts on a wide scale throughout long time-frames – that is, at the local, mainland or even worldwide scale over several years [8].

Distant detecting can provide reciprocal data to existing ground-based natural checking frameworks. It may very well be utilized to address the issue for ideal data and can give brief cross-limit data. Information and data acquired through earth perception can be utilized inside geographic data frameworks for overlay and correlation with other geo-referred to data [9].

## 4.2.6 Enterprise Monitoring

Enterprise ecological observing and reporting is the arrangement of measures executed and paid for by administrators, the normal or lawful people practicing real control over the specialized working of the office. Such a framework incorporates persistent and occasional perceptions, the account, stockpiling and treatment of information identifying with ecological assurance and the revealing of the outcomes to the administration and every one of the workers of ventures, the public specialists and the overall population as sets of essential, determined or totaled information and general data [10].

Reinforcing the endeavor of ecological observing and reporting will further develop the checking of big business in line with guidelines. Expanding the amount of ecological data created by undertakings, working on the nature of this data and improving access to it for the overall population will assist with applying critical tension on polluters to reduce their unfriendly ecological effect [11].

## 4.3 WSN FOR ENVIRONMENTAL MONITORING

Improvement in the innovation of sensors like Micro Electro Mechanical Systems, remote interchanges, installed frameworks, conveyed preparing and remote sensor applications have recently contributed an enormous change in Wireless Sensor Network (WSN). This helps and further develops work execution both in industry and in our every day lives [12]. Remote sensor networks have been broadly utilized in numerous spaces, particularly for observation and checking in horticulture and territory checking. Climate checking has become a significant field of control and assurance, giving ongoing framework and control correspondence with the actual world. An insightful and savvy WSN framework can accumulate and deal with a lot of information from the start of observing, and can oversee air quality, the state of traffic, and climate circumstances. This chapter examines and surveys remote sensor network applications for natural checking [13]. To carry out a decent observing framework, there are a few prerequisites to be followed [14].It is demonstrated that these methodologies can further develop the framework execution, give a helpful and effective strategy, and can likewise satisfy utilitarian prerequisites. Recent advancements in remote correspondences and hardware have brought the vision of WSN into reality, which has expanded the development at minimal expense, low force and multi-practical sensors that are less costly and have good availability. Every hub comprises of microcontrollers, memory and handset [15]. The microcontrollers are utilized to execute tasks, prepare information and help the usefulness of different segments in the sensor hub. For the memory, it is principally utilized for information stockpiling, while the handset acts from the mix of transmitter and beneficiary capacities [1]. Information like temperature, light and sound are gathered by sensors and afterward sent to a worker. These battery fueled hubs are utilized to screen and control the actual climate from distant areas [16]. In the previous few years, WSNs have been generally utilized and applied in clinical, military, modern, farming and ecological checking.

Natural checking has been a significant part of WSN applications. It develops generally along with the advancement of ongoing innovation [17]. By and large, ecological checking framework controls and screens climate boundaries like temperature, stickiness, light and pressing factor. There are a few examinations that concentrate in ecological checking applications [4] [5]. A few specialists execute the deficiency open minded and studies the tradeoff between device cost and lifetime of sensor network [6] to ensure the adaptation to non-critical failure is in the three dimensional settings. [7] and [8] created multi jump correspondence applications, which implies the information of the temperature and dampness will be communicated to the neighbor hub and afterward shipped off the end client PC [18]. The ecological boundary information estimation will show the outcome utilizing Java [3] and the information is deciphered into a chart and table. Accordingly, it is important to comprehend the necessities for the improvement of checking applications [9].

### 4.3.1 Autonomy
In view of the fact that the radio handset uses high energy consumption, it is necessary to ensure the battery utilized can work sufficiently throughout the process, and the set-up should be energy-efficient [19].

### 4.3.2 Reliability
Basic maintenance and unsurprising activities are required to avoid unforeseen accidents. Additionally, upkeep by any individual ought to be avoided in light of the fact that n wireless sensor network the end client may face trouble in getting information in time during the terrible climate condition [20].

### 4.3.3 Robustness

The organization must be powerful enough to cope with issues like equipment failure and a weak signal network. For instance, the impact of dampness can bring short-out issue and lead to framework reboot [21].

## 4.4 CLIMATE MONITORING SYSTEM APPLICATIONS

As of late, the advancement of ecological observing frameworks has been applied in numerous applications to help individuals in their work and to diminish cost and time [22].The uses of natural observing have quickly been taken up in rural observing, natural surroundings checking, indoor checking, nursery checking, environment checking and timberland checking. It brings the advantage of the local population understanding the significance of the remote sensor network innovations in their day-to-day existence [23].

## 4.5 AGRICULTURAL MONITORING

Rural observing consistently centers for the most part around the area of cultivation [24].A few examinations characterize creature observing as creature following [10] however the idea is something very similar as explained in Figure 4.1. There are techniques to be carried out to get past each stages well defined for the whole life cycle [11]. The cooperation among creatures and human has been created and perceived for many years. The commitment of creatures love, genuine hearted and congruity live can have a positive effect on human physical and mental health [12, 13].Be that as it may, these days numerous creatures need appropriate treatment and there are additionally situations where these creatures' illnesses are not recognized. Subsequently, it is important to have a checking framework to screen creature behaviour and produce a report with respect to their well-being or conduct framework progressively [25].

There are numerous distinguishing techniques in checking creature well-being, yet some of them either come up short or are ailing in proficiency or ease of use [26]. The plan of RFID-based Mobile Monitoring System (RFID-MMS) [14] helps clients control creature behavior and development. [15] and [16] proposed following requirements for wild creature checking. It will screen the living space, and provide examples of development and creature behavior. Wild lynxes or canines are utilized as the objective subjects. The sensor hubs positioned around the collars will gather

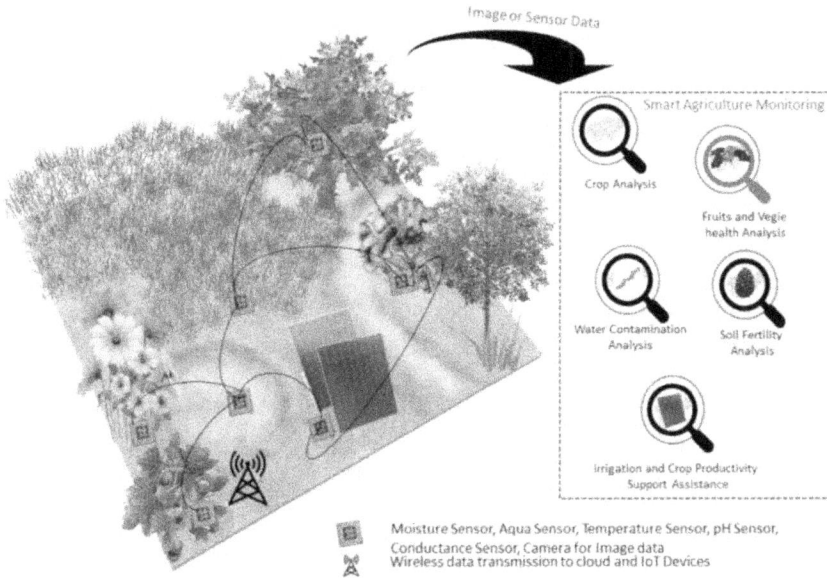

FIGURE 4.1    Agriculture monitoring using WSN.

GPS data and information of multimodal sensor, appropriate through the framework to the customer. From tests done, it shows that the scope of sign correspondence can be accomplished from 200–250 meters and this ought to be in thought all together to plan a self-supportable framework which is more productive later on [17]. A rural climate observing framework which incorporates the sensor hubs plan equipment and programming improvement which comprises of the product flowchart was assembled. From the test carried out, the framework demonstrated low energy consumption but high dependability, which can handle constant checking for unprotected farming and natural observing. The observing framework for poultry likewise contributes a major benefit to clients, particularly ranchers [18].Proposed and fostered a poultry observing framework which is online application. They utilize Crossbow's TelosB bits that can incorporate with the sensors to gauge the temperature and mugginess of the chicken [27]. Toward the finish of the investigation, they got most extreme distance of signal reach up to 40 meter with 5% bundle misfortune mediocre. From the outcome, they have presumed that the framework is able to recognize climate irregularities in the chicken homestead. This sort of checking isn't just applied for poultry, but also for dairy cattle checking [19, 20].

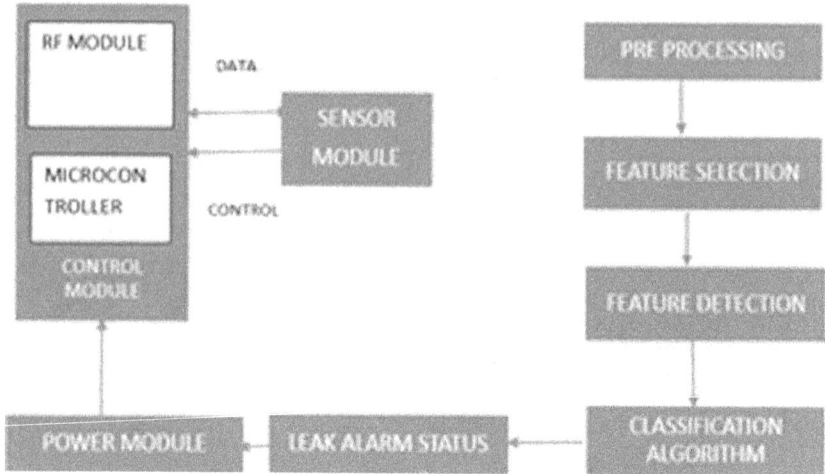

FIGURE 4.2    Weather monitoring using WSN.

## 4.6  HABITAT MONITORING

Natural surroundings checking is one of the fundamental parts of eco-logical observing [28]. Territory implies a spot where a creature or plant normally develops or lives. Thus, living space checking is essential to ensure their species' autonomy and to forestall any environmental aggravation for creatures and plants. Contamination can create adverse consequences for wellbeing and biological equilibrium [29]. Accordingly, a framework is needed that can screen contamination so it is kept under control [21], and a web-based graphical user interface (UI) to deal with the information of contamination productively. The sensor hubs are utilized to view current sensor monitoring. Toward the end of the examination, they figure out how to work on the presentation of the sensors innovation by acquiring a stable correspondence despite the fact that the normal lifetime of the sensors has declined because of the necessity of inertness [30]. Figure 4.2 shows the weather monitoring system.

## 4.7  ARTIFICIAL INTELLIGENCE AND WSN

While the principle objective of artificial intelligence (AI) is to foster frameworks that imitate the scholarly and communication capacities of an individual, distributed artificial intelligence (DAI) seeks a similar goal but zeroes in person social orders. A worldview in current use for the improvement of DAI depends on the thought of multi-specialist frameworks [31]. A multi-specialist framework is shaped by various collaborating clever frameworks called specialists, and can be executed as a product program, as a committed PC, or as a robot. Insightful specialists in a multi-specialist framework interface among one another to put together their design, dole out errands, and exchange information [32]. Ideas identified with multi-specialist frameworks, fake social orders, and mimicked associations give rise to another worldview, which includes issues such as participation and rivalry, coordination, cooperation, correspondence and language conventions, arrangement, agreement advancement, struggle discovery and goal, aggregate knowledge exercises led by specialists (e.g. issue goal, arranging, learning, and dynamic in a dispersed way), intellectual different insight exercises, social and dynamic organizing, decentralized organization and control, security, dependability, and heartiness (administration quality boundaries) [33]. Circulated astute sensor organizations can be seen according to the point of view of a framework created by various specialists (sensor hubs), with sensors working among themselves and shaping an aggregate framework whose capacity is to gather information from actual factors of frameworks. Consequently, sensor organizations can be viewed as multi-specialist frameworks or as counterfeit coordinated social orders that can see their current circumstance through sensors. However, the inquiry is how to execute AI instruments inside WSNs. There are two potential ways to deal with the issue: as indicated by the main methodology, planners have as a primary concern of the worldwide target to be cultivated and plan both – the specialists and the communication component of the multi-specialist framework [34]. In the subsequent methodology, the originator imagines and develops a bunch of self-intrigued specialists who then, at that point, advance and interface in a steady way, in their design, through developmental procedures for learning. A similar issue applies when working with a WSN viewpoint seen from the perspective of DAI. Will the standards, calculations and use of DAI be utilized to enhance an organization of dispersed remote sensors? Is it conceivable to execute an answer that empowers a sensor organization to act as a wise multi-specialist framework? According to the point of view of multi-specialists, counterfeit social orders, and mimicked associations, how should a conveyed sensor network be introduced in a proficient way and accomplish the proposed goals of taking proportions of actual factors without anyone else? What are the association focuses between DAI and wireless sensor organizations? The major thought in this section is to propose a model that empowers an exceptionally dispersed sensor organization to act cleverly as a multi-specialist framework [35].

## 4.8 REMOTE SENSOR NETWORKS

A sensor network (SN) is a framework that comprises thousands of little stations called sensor hubs. The principle function of sensor hubs it is to screen, record and report a particular condition at different areas to different stations [36]. Likewise, an SN is a gathering of particular transducers with a correspondences framework proposed to screen and record conditions at assorted areas. Categories normally checked are: temperature, stickiness, pressure, wind course and speed, brightening force, vibration power, sound power, power-line voltage, substance focuses, contamination levels and fundamental body capacities. Sensor hubs can be envisioned as little PCs, amazingly essential as far as their interfaces and their parts are concerned [37]. Albeit these gadgets have almost no ability on their own, they have significant preparing capacities when they are filling in as a total very hub in a sensor network is normally furnished with a radio handset or other remote specialized gadget, a little microcontroller, and a fuel source, generally a battery. A sensor hub may fluctuate in size from that of a shoebox down to the size of a grain of residue [7]. A sensor network regularly establishes a remote specially appointed organization, implying that every sensor upholds a multi-jump steering calculation (a few hubs may advance information bundles to the base station). Note that SNs have a much greater energy requirement than PDAs, cell phones or workstations [38]. The entire structure is ordinarily under the organization of one regulator: the base station. The primary usefulness of the base station is as passage to another organization, and has amazing information processor and capacity focus. Advances in microelectronics and remote correspondences have made WSNs the anticipated panacea for tackling a large group of enormous scope choice and data handling errands. The applications for WSNs fluctuate, commonly including some sort of observing, following or controlling. Explicit applications incorporate natural surroundings checking, object following, atomic reactor control, fire recognition and traffic observing [39]. In a commonplace application, a WSN is dispersed in a locale where it is intended to gather information through its sensor hubs. Various WSNs have been conveyed for natural checking [40]. Remote sensor networks have been produced for apparatus condition-based maintenance since they offer massive expense reserve funds and empower new functionalities. Although various new WSN frameworks and advances have been created, new issues or difficulties are yet to be settled or resolved. Such issues include: ideal steering systems, life expectancy of the WSN, lifetime of the hubs being extremely restricted, re-configurability without redeployment. Finally, since WSNs become famous there is definitely not a typical stage. Some delegate plans have more extensive clients and engineer networks, like Berkeley Motes, which was the principal business bits stage [41].

## 4.9 HUMAN-MADE CONSCIOUSNESS AND MULTI-AGENT SYSTEMS

Classical AI pointed toward copying inside PCs the scholarly and collaboration capacities of an individual [42]. The advanced way to deal with AI revolves around the idea of a normal specialist. A specialist is whatever can see its current circumstance through sensors and follow up on that climate through actuators. A specialist that consistently attempts to enhance a proper presentation measure is known as a judicious specialist. The meaning of a reasonable specialist is genuinely broad and can incorporate human specialists (having eyes as sensors, hands as actuators), mechanical specialists (having cameras as sensors, wheels as actuators), or programming specialists (having a graphical UI as sensor and as actuator). According to this point of view, AI can be viewed as the investigation of the standards and plan of counterfeit objective specialists [43]. Be that as it may, specialists are only occasionally independent frameworks. Much of the time they coincide and collaborate with different specialists in a few distinct manners. Models incorporate clever web programming specialists, soccer playing robots, internet business arranging specialists, PC vision devoted specialists, and more. Such a framework comprising a gathering of specialists that can conceivably connect with one another is known as a multi-agent systems, and the comparing subfield of AI f multi-specialist frameworks is distributed AI (DAI).

### 4.9.1 Remote Sensor Networks and AI

A savvy sensor is one that changes its inner conduct to upgrade its capacity to gather information from the actual world and imparts it in a responsive way, to a base station or to a host framework. The usefulness of a keen sensor incorporates: self-alignment, self-approval and remuneration. The self-adjustment implies that the sensor can screen the estimating condition to choose whether another alignment is required or not. [44]. The sorts of human-made reasoning procedures generally utilized in enterprises are: artificial neural network (ANN), fuzzy logic and neuro-fuzzy. Savvy sensor structures implanted in WSNs bring about remote smart sensors [45]. The utilization of artificial knowledge methods assumes a vital part in building insightful sensor structures. Primary examination issues of the WSNs are centered around the inclusion, availability network lifetime, and information loyalty. In recent years, there has been an expanding interest in the space of AI and DAI, and their techniques for tackling WSNs compels, make new calculations and new applications for WSNs [46].Across the board is a fundamental element of a middleware answer for WSN. This incorporates beginning sensor-determination and assignment designation as runtime variation of distributed errand/assets [47]. The boundaries to be upgraded incorporate energy, transfer speed and organization lifetime. In this standard particular case distributed independent reinforcement learning proposed the utilization of aggregate insight in asset the executives inside WSNs. At long last, smart systems' administration and community frameworks are likewise proposed as WSN improvement [48].

Remote sensor organizations (WSNs) are a growing innovation because of late headway in extremely limited scope manufacturability and high-scale reconciliation of different electronic parts in a solitary bundling. A nal sensor hub (or bit) is an independent bundle of hardware important for holding various sensors, an implanted microcontroller, a force unit that has restricted limit, which might possibly be inexhaustible, and a radio trans receiver at its center. The typical size of a sensor hub can be anywhere from a matchbox to a coin, yet it is expected to shrink significantly in the following decade with the astonishing guarantee of nanotechnology assembling and manufacture. Current and extended utilization of remote sensor networks incorporates a wide assortment of spaces, which have been generally difficult to access for numerous reasons, including likely damage to people, being at far-off locales or being circulated over extremely large regions, and being liable to unforgiving geo or meteorological conditions, among others [49]. Observing climate (contamination in a lake or stream), timberland fires, volcanoes, combat zone troop developments, human body, checking underlying wellbeing of elevated structures or extensions, brilliant home computerization, and last but not least, keen sustainable power matrix checking and control are among the endless possible applications [50].

Given the idea of utilizations for remote sensor organizations, the normal arrangement situation, as a rule, involves the dispersed arbitrary position of hundreds or thousands of such hubs in a given geographic region through either dropping loftily from a flying specialty or spreading from a moving area vehicle. Resultantly, the arrangement of sensor hubs structure an impromptu remote PC organization. Frequently such hubs and the organization are relied upon to work for a time of somewhere around one to two years utilizing the on-board power source contingent upon the idea of the application with no external upkeep or repair access since such access may just not be feasible [51]. Remote sensor networks are imagined to be sent and expected to work independently for various years especially in non-accommodating conditions without human inclusion. Different components, including topography, environment and human-instigated purposeful or non-deliberate impedance in the electromagnetic range will antagonistically influence the conveyed network, and henceforth require a decent degree of flexibility to evolving conditions. A remote sensor network is a unique framework as in it goes through changes after some time, which have significant outcomes on the activity and prerequisites of the organization [52].

A portion of these progressions may incorporate updates to mission or usefulness at various scales; changes in static or dynamic hub sythesis; energy utilization profile of hubs and the organization over the long run; obliteration or passing of specific hubs; and transient impacts that may briefly upset a hub, a group of hubs, or a sub-organization to work inside its typical working system. There are a number of between related enhancement measures, for example least energy, information misfortune, unwavering quality, strength, and so on, set up during the plan and activity of remote sensor organizations. In the ordinary plan and advancement for remote sensor organizations, a particular arrangement of conventions for medium access, limitation and situating, time synchronization, geography control, security and steering are distinguished dependent on the current design of the organization, the prerequisites of the application and the geography of their sending [53].

Notwithstanding, unexpected behavior might be possible for a wide range of reasons following organization, like unexpected demise of sensor hubs, unsuitable execution of utilization rationale, geography changes and transformed organization conditions. For example, it is possible that versatile convention choice or exchanging plans might be created, which may react all the more ideally to changes that influence the remote sensor network over the long haul [54]. This prompts the need to change the product behavior at both the convention and application layers after the organization has been sent. The objective of the venture, revealed in this chapter, remote sensor networks for altogether improved self-ruling conduct and activity [55]. The assumption from a versatile and keen WSN is that it can promptly think about the progressions directed by the unique idea of functional and application perspectives, and in like manner adjust to evolving conditions, mission, and functional requests following the sending [56]. The helpful variation capacity can be presented through installing a counterfeit neural organization, which can be launched to any specific structure, for example, feed-forward, self-construction or repetitive, in completely equal and circulated mode inside the remote sensor organization.

A neural organization works into remote sensor networks in equal and dispersed calculation mode [57]. Remote sensor organizations (WSNs) are topologically like fake neural organizations. A WSN is established from hundreds or thousands of sensor hubs or bits, every one of which normally has generous computational force (through the locally available microcontroller) [58]. A counterfeit neural organization (ANN) is made out of hundreds or thousands of (computational) hubs or neurons, every one of which is accepted to have or require extremely restricted computational handling capacity [59]. This topological comparability can be the premise to profit the transformation and functional parts of WSNs through utilizing the current neural organization hypothesis completely overall. Since a remote sensor network with a huge number of bits or hubs is a dispersed framework with equal registering capacity, a combination with another equal and disseminated framework, the counterfeit neural organization is a characteristic result. Truth be told, there is balanced correspondence, in that a sensor bit can behave like or execute a neural organization neuron or hub, while remote connections among the bits are comparable to the weighted associations among neurons [60].

Melding WSNs with ANNs makes way for WSNs unexpectedly to have extensively generous computational insight to have the option to address an extremely complete arrangement of issues both at the convention and application layers. The computational force of fake neural organizations is all around recorded in the writing and for a far reaching range of issue spaces [61]. Arrangement, work estimation or relapse, advancement, grouping, framework distinguishing proof, and expectation (e.g. time series) are among the main applications. Counterfeit neural organizations have arisen as a commonsense innovation for work guess dependent on a huge arrangement of accessible examplar designs [1].

Fake neural organizations (ANN) are general approximators that can be prepared on an informational index to plan multi-dimensional nonlinear capacities [2]. Indeed, ANNs offer an incredible and general system for addressing non-direct mappings from multi-contribution to multi-yield factors, where various flexible boundaries control the type of the planning [3]. ANNs are especially alluring since quite a bit of univariate guess approaches neglect to sum up well to higher dimensional spaces: splines and wavelets perform well in relapse and sign examination if the dimensionality of the information space is close to three [4–6]. There is a considerable group of information relating to the capacity estimate abilities of counterfeit neural organizations. It was hypothetically settled (alongside generous observational proof) that a one-covered up layer feed-forward network, whose neuron yield capacities are sigmoidal (e.g. neuron input is planned to yield through a sigmoid shaped work), is fit for approximating self-assertive (nonstop) work [1,7–9].

Feedforward neural organizations including multi-facet perceptron (MLP) and spiral premise work (RBF) are notable for characterization and capacity estimate, while intermittent neural organizations like the time-defer neural organizations or Elman neural organizations are helpful for time series determining or forecast. For enhancement and acquainted memory applications, Hopfield-style networks including mean-field strengthening and Boltzman machine are utilized. There are ANN calculations fitting for (unaided or semi-managed) grouping applications – putting together guide and its numerous variations like the group of versatile reverberation hypothesis (craftsmanship) neural organizations and straight vector quantizer among numerous others [62]. All in all, ANNs can resolve numerous central issues in figuring and as such offer a conventional processing instrument that is profoundly flexible, and has significant utility for a thorough arrangement of issue areas [63].

Properties of the proposed WSN-ANN design for adaptability and intricacy are of basic interest. In particular, these characteristics incorporate the capacity of the WSN-ANN to scale with the issue size, the computational intricacy in reality, and the interchanges or informing intricacy [64]. It is pertinent to note that computational intricacy parts of neural organizations is a space that is, to a great extent, inadequate and divided despite the fact that there have been essential advances during the last decade [10–15]. There are such a large number of various neural organization standards and incalculable boundaries to consider for a combined and cognizant treatment of the subject, which in this way prompted just a predetermined number of computational intricacy investigations for explicit occasions of neural organization calculations and related learning or preparing measures. The time intricacy of the proposed registering framework is dictated by various elements relying upon the kind of neural organization [65].

There are commonly two particular stages that should be considered: preparing the neural organization, which sometimes bears a generous time cost, and the subsequent arrangement whose time cost will in general be insignificant compared with that of the preparation. For instance, for feed-forward multi-facet neural organizations, the preparation time is principally directed by the intermingling properties of the particular issue being tended to, which additionally influences the geography of the neural organization. The assembly properties of multi-facet feed-forward networks differ drastically starting with one issue area then onto the next [66]. The experimentally indicated intermingling measure, for example one being combined mistake fulfilling a client characterized upper bound, additionally assumes a critical part in the time intricacy. There will likewise be ready preparing time related with carrying out the neuron elements signal handling which is insignificant contrasted with other expense components, and consequently may and will be overlooked for the remainder of the conversation. These expenses are now intrinsic in the neural organization calculation paying little heed to its equipment acknowledgment at a particular stage [67]. There is, in any case, another expense part due to the execution of the neural organization calculation on a remote sensor organization. It is normal that there will be delays infused into the learning or preparing measure because of the need to trade neuron yield esteems among the bits (neurons) through the remote medium as overseen by a fitting medium access control (MAC) convention since MACs will fundamentally happen and must be managed. This means the MAC convention and the informing prerequisites of a particular neural organization as shown by its between neuron availability ascribes will assume a part in the finish of the time cost appraisal. Hence, it is sensible to propose that the time intricacy will fluctuate drastically dependent on the issue area, the organization type and design, boundary settings and other variables [68].

A far-off sensor organize (WSN) contains close to nothing, low-controlled sensors talking with each other maybe through multihop distant associations and collaborating to accomplish a run of the mill task. A far-off sensor coordinate is a plan of close to nothing, distantly bestowing centers where each center point is equipped with various parts [5]. The center points grant distantly and routinely self-mastermind ensuing to being passed on in an improved structure [69].Such a framework is envisioned to arrange the actual world with the web and computations. The power supply at each center is, by and large, limited, and replacement of the batteries is a significant and limiting part of the time as a result of the enormous number of the center points in the framework. Each center may contain various kinds of memory (program, data and burst memories), getting ready limit (something like one microcontrollers, computer processors or Digital Signal Processor chips), have a RF handset (generally with a singular omnidirectional radio wire), have a force source (e.g., batteries and sun situated cells), and suit various sensors and actuators. Sensor centers collaborate with each other to perform endeavors of data recognizing, data correspondence, and data taking care of [2]. Structures of 1000s or even 10,000 centers are predicted. Such structures can change the way wherein we live and work. Advances in sensor development and far-off exchanges have allowed planning and improvement of huge scope and viable sensor arranges that are fitting for various applications, for instance, prosperity noticing, regular checking and disaster area perception. A vital point in the arrangement of WSNs is to keep them valuable for whatever time allotment is possible [70]. Because of the low energy (or imperativeness) of the battery, the sensors can absolutely deplete the essentialness or have extra essentialness under the cut-off needed for the sensors to work accurately. These sensors are called broken considering the way that they can't play out any checking tasks accurately. It is said that a WSN is helpful if at whatever point there is something like one correspondence way between each consolidate of non-broken sensors in the framework [71]. Regardless, the presence of correspondence ways between sensor sets is related to another essential property of the WSN, called vertex accessibility (or fundamentally network). All things considered, recognizable proof applications should be fault lenient, in which any match of sensors is, by and large, related by different correspondence channels. Consequently, the framework handiness and, subsequently, the variation to inward disappointment of the framework depend to a large degree on accessibility.

Far-off sensor frameworks can recognize and advance recognized data and perform responses reliant upon the obtained data in an appropriate manner. The WSN contains sensor centers and sink centers. Sensor centers generally have low expenses, obliged power and limited transmission broaden; they are responsible for recognizing events or distinguishing normal data. Sink centers are more resource rich center points with bountiful wellsprings of imperativeness, more vital correspondence and computational cut-off, and the ability to perform incredible reactions. Exactly when the tolerant center plays out a movement, these centers are called performing craftsman centers. Right when a sensor center distinguishes a couple of data that will be passed on to its checking an area, it will communicate the event to adjoining center points, which along these lines will send the event back to another skip. At the point when the tolerant center gets the data, it will play out the relating reactions fittingly. WSNs license some sensible applications, for instance, military control, ponders control and attack revelation [1].

As of now, far-off sensor frameworks are starting to increase at an animated rate. It isn't unrealistic to expect that in the following 10–15 years, the world will be populated with distance sensor sorts that can be obtained through the web. This could be practically identical to the web transforming into an actual framework. This new development is invigorating, with endless potential for certain zones of usage, including regular, remedial, military, transportation, public boundary and smart spaces. Since a distance sensor puts together is an appropriated structure ceaselessly, a trademark request is the number of nodes are dispersed logically in these new structures. Unfortunately, not a great numerous previous occupations can be associated and new plans of deployment of nodes are needed in each part of the structure. The majority of assessments of appropriated structures in the past have expected that the systems are wired, have limited power, are not steady, have UIs, for instance, screens and mice, have a settled plan of resources, treat every center point in the structure as fundamental and are not restricted by position. On the other hand, for far-off sensor frameworks, systems are distant, have low energy requirements, are persistent, use sensors and actuators as interfaces, have logically changing resource sets, all out lead is objective and position is essential. Various far-off sensor coordinates in like manner use devices with unimportant breaking point, which puts a further weight on the probability of using past plans. Disregarding the way that sensor frameworks are an extraordinary kind of exceptionally delegated frameworks, shows plans for extemporaneous frameworks can't be used as they are for sensor orchestrates as a result of the following reasons: 1) The quantity of hubs in the sensor networks is extremely large and should scale a greater number of significant degrees than specially appointed organizations and thus requires unique and more adaptable arrangements. 2) The information transmission rate is required to be exceptionally low in WSN and is factual in nature. Be that as it may, the specially appointed versatile organization (MANET) is intended to convey rich sight and sound information and is principally carried out for appropriated registering. 3) A solitary sensor network is normally carried out by a solitary proprietor; however MANET is typically performed by various autonomous elements [4]. 4) The sensor networks are information driven, that is, the questions in the sensor network are coordinated to hubs that have information that meet certain conditions and a solitary tending to is unimaginable, as they don't have worldwide identifiers. Be that as it may, MANET is fixated on hubs, with questions addressed to specific hubs determined by their novel locations. 5) The sensor hubs are ordinarily executed once in their valuable life and those hubs are generally fixed, except for some portable hubs, while the hubs in MANET move in a specially appointed way. 6) Like the hubs of the MANET sensors, they are additionally intended for auto design, yet the distinction in rush hour gridlock and energy utilization requires separate arrangements. Concerning impromptu organizations, sensor hubs have restricted force supply and energy re-energizing isn't reasonable, considering the huge number of hubs and the climate where they are carried out.

Sensor networks share normal mistake issues (like association blunders and clog) with conventional wired and remote conveyed networks, just as presenting new wellsprings of deficiencies (like hub disappointments). Adaptation to non-critical failure strategies for disseminated frameworks incorporate instruments that have become industry norms like simple network management protocol and Transmission Control Protocol/Internet Protocol(TCP/IP), just as more specific and/or more proficient techniques that have been broadly considered [14]. Shortcomings in sensor organizations can not be settled similarly as customary wired or remote organizations because of the following reasons: 1) Conventional organization protocols, by and large, couldn't care less about energy utilization, as wired organizations have steady and specially appointed power, and remote gadgets can be re-energized intermittently; 2) Conventional organization protocols mean to accomplish a solid highlight point, while remote sensor networks concur with dependable occasion discovery; 3) In sensor organizations, blunder hubs happen more frequently than in wired organizations, where it is accepted that workers, switches and customer machines normally work more often than not; This suggests that a nearer hub wellbeing checking framework is required without causing huge overheads; 4) Conventional remote organization protocols are dependent on utilitarian level conventions to keep away from bundle impacts, the secret terminal issue and station mistakes utilizing the administrator's actual sense and sense of the virtual administrator (station observing). Numerous identification calculations are inexactly characterized deformity examples or disappointments too broad definition [6]. Controvercial distinguishing cross-over in line deserts dependent on wide meanings of approval mistakes. Looking past the strategies of location and amendment of mistakes, there has been critical work that outlines provide a scientific classification deficiency.

Scientific classification of shortcoming open minded strategies late examination has fostered a few methods that arrangement with various sorts of mistakes in various layers of the organization stack [72]. To assist with understanding the theories, the methodology and the bits of knowledge behind the plan and improvement of these strategies, the scientific classification of the diverse adaptation to internal failure procedures utilized in conventional appropriated frameworks [15] was given as: 1) Disappointment avoidance: to foresee or forestall disappointments; 2) Recognition of issues: this involves utilizing various measurements to gather the manifestations of potential blunders; 3) Deficiency detachment: this is to connect various sorts of shortcoming signs (alerts) got from the organization and to propose various speculations of disappointment; 4) Issue ID: this involves testing every one of the theories proposed to find and precisely recognize issues. 5) Shortcoming recuperation: this is to manage disappointments, that is, to switch their adverse consequences. The recognizable proof and confinement of flaws are here and there, by and large, characterized as deficiency diagnostics. As a general rule, these procedures work at various levels of the organization convention stack. Most disappointment anticipation methods work at the organization level, adding repetition in directing courses; most shortcoming recognition and recuperation procedures work in the vehicle layer; and some mistake recuperation strategies are acted in the application layer, concealing disappointments during the web information preparing.

Disappointment Identification Approaches in the WSN. The brought together methodology is a typical answer for distinguishing and recognizing the reason for suspected mistakes or hubs in WSN. Generally, a geologically or intelligently brought together sensor hub (as far as base station [5, 17, 18], focal regulator or head [4], sink) accepts accountability for observing and finding blemished or resistant hubs in the organization. The vast majority of these methodologies consider that the focal hub has limitless assets (for instance, energy) and can play out a wide scope of mistake taking care of upkeep. They additionally accept that the helpful existence of the organization can be broadened if the mind boggling organization works and the transmission of messages can be changed to the focal hub. The focal hub, as a rule, embraces a functioning location model to reestablish network execution states and individual sensor hubs by occasionally infusing solicitations (or questions) into the organization, and dissecting this data to recognize and find flawed or suspected hubs. In [17], the base station utilizes checked bundles (containing geological data of beginning and objective areas, and so on) to distinguish sensors. It depends on hub reaction to distinguish and disengage suspect hubs in steering ways when an over the top bundle fall is recognized or compromised information is identified. Also, the focal chairman gives an incorporated way to deal with forestalling likely disappointments by contrasting the current or recorded conditions of the sensor hubs with the overall data models

Given the idea of utilizations for remote sensor organizations, commonplace sending situation much of the time involves dispersed arbitrary arrangement of hundreds or thousands of such hubs in a given geographic region through either dropping loftily from a flying art or spreading from a moving area vehicle. Resultantly, the arrangement of sensor hubs structure an impromptu remote PC organization. Frequently such hubs and the organization are required to work for a time of something like one to two years utilizing the on-board power source contingent upon the idea of the application with no external upkeep or fix access since such access may basically be not viable. Remote sensor networks are considered to be sent and expected to work self-governingly for various years especially in non-affable conditions without human inclusion. Different variables including topography, environment and human-actuated deliberate or non-purposeful obstruction in the electromagnetic range will unfavorably influence the sent organization, and thus require a decent degree of versatility to evolving conditions. A remote sensor network is a powerful framework as it goes through changes over the long term, which have significant results on the activity and prerequisites of the organization. A portion of these progressions may incorporate modifications to mission or usefulness at various scales; changes in static or dynamic hub creation; energy utilization profile of hubs and the organization after some time; obliteration or passing of specific hubs; and transient impacts that may briefly upset a hub, a group of hubs, or a sub-organization to work inside its ordinary working structure. There are a bunch of between related improvement measures, for example least energy, information misfortune, dependability, vigor, and so forth, set up during the plan and activity of remote sensor organizations. In the run of the mill plan and improvement for remote sensor organizations, a particular arrangement of conventions for medium access, limitation and situating, time synchronization, geography control, security and directing are recognized dependent on the current set-up of the organization, the prerequisites of the application and the geography of their sending. Nonetheless, lackluster showing or startling conduct might be possible for a wide range of reasons following arrangement, like unexpected passing of sensor hubs, unsuitable execution of utilization rationale, geography changes and transformed organization conditions. For example, it is possible that versatile convention choice or exchanging plans might be created, which may react all the more ideally to changes that influence the remote sensor network over the long term. This prompts the need of changing the product conduct at both the convention and application layers after the organization has been conveyed [73].

An Artificial Neural Network (ANN) neurons passage is a product program application or real framework that fills in as the association factor between the cloud and regulators. Sensors unit demonstrated in this network providing a Gateway as an interpreter, which permit two various conventions to talk and steering the records to Cloud stage. Entire insights motile to the cloud passes through the passage and sensors are a committed equipment gear or programming program. An ANN entryway may furthermore be alluded to as an insightful passage or adapt to level [74].

However, in local area the thousands or more loads of nodes are connected, but the dependency algorithm of the gadget (artificial machineries) is the principal task. The conventional prerequisite adjustment is utilize in this situation. These are low profile conventions in expressions of data transmission, battery controlled speed. In ANN, no possibility to change the measurements of Cloud stage by using dependency algorithm. [75]

ANN statistics storehouse is a database improved to dissect relational statistics coming from transactional structures and line of enterprise operations. A statistics lake in ANN, is a centralized depository to protect all structured and unshaped data at any scale [76].

Safety is one of the primary rudiments of ANN. The ANN units are regularly accumulating extraordinarily non-public data, and bringing the factual world for analysis. Connected sensors produce massive volumes of data, which want to be securely transmitted and protected from cyber-criminals [77].

When the connection and dexterity of new fashions are examined and accredited by means of data, a new establishing algorithm is used through neural operations [78]. The instructions despatcher with managing applications use to save a large records in storehouse. This may also help interrogate into complex cases [79].

An ANN installed inside a WSN, which can be viewed as a universally useful PC that additionally turns out to be greatly equal and easily circulated, offers tremendous potential to perform calculations of conventional nature and high utility. Truth be told, the WSN-ANN configuration can tackle huge scope issues progressively because of its completely equal and enormously conveyed engineering. A non-comprehensive rundown covers issues from the spaces of straight, quadratic, and direct complementarity issues; frameworks of direct conditions; least squares issue; minimax (Linfinity-standard) arrangement of over-resolved arrangement of (direct) conditions, and least total deviation (L1-standard) arrangement of frameworks of conditions; discrete Fourier change; lattice variable based math issues including reversal, LU deterioration, QR factorization, unearthly factorization, SVD, Lyapuno assessment; chart hypothetical issues; and static advancement issues [80,81].

Most studies of remote sensor organizations/networks (WSNs) are pointed toward working on such highlights as proficiency, unwavering quality and security of the framework. This chapter centers around further developing the energy proficiency of WSNs. Since a large portion of the hubs work independently, the idea of the utilization of energy assets straightforwardly influences the existence pattern of the organization. A large chunk of the energy of organization hubs is devoured by the cycle of correspondence through remote correspondence. A basic answer for reducing the quantity of correspondence creates a specialized logical inconsistency: "With an expansion in the quantity of correspondence meetings, the energy utilization of hubs expands, which prompts a decrease in the hour of self-governing activity of the hub; with an abatement" – the likelihood of missing significant occasions builds, which may prompt erroneous dynamic [82].

*Energy-productivity for information transmission between the components of the sensor network, control module settles on a choice about the condition of the sensor network dependent on the arrangement of the information communicated from the sensors. The way toward registering the organization state utilizing the choice tree is idempotent, that is, if similar information is given, the outcome is indistinguishable. Subsequently, assuming the information, communicated from the sensors, have not changed, the organization state stays as before [70]. The computation of the organization state is done through the substitute correlation of the readings of the sensors with the predicates got before in the preparation of the model. The examination proceeds until a leaf of the tree with the worth of the specific condition of the framework is accomplished [71]. Many calculations for energy productive information transmission is that every sensor sends information just when it can impact the choice of the principle module. Vertical predicates structure mathematical scopes of qualities, inside which the condition of the organization will be unaltered. Hence, in the event that the current sensor perusing is inside this reach, the information may not be sent. For this situation, the control module ought to get that assuming a specific sensor doesn't communicate information, its readings have not changed essentially and it is feasible to utilize the past esteem. Because of the use of this calculation, the computational burden is dispersed:* before the control module settles on a choice about the condition of the framework, every sensor freely settles on a choice whether its present status is significant when the primary module makes a choice [74].

*Remote Sensor Network (WSN) alludes to an organization of little, low-power sensing points that can detect and impart data about their current circumstance. WSNs are possible for different applications with various spatial organizations that reach from being meager to exceptionally thick by getting implanted in actual conditions. The batterypowered remote sensor hubs are equipped for various kinds of data, gain from this data and settle on choices progressively. These frameworks share similar objective of distinguishing fascinating occasions with regards to an obscure climate throughout some stretch of time. New remote specialized gadgets have attracted consideration because of innovations like detecting abilities, correspondence conventions and energy reaping and implanted frameworks. Our methodology empowers the sensor organization to utilize AI (ML) strategies to report the focal control about irregularity location and its force self-governing. A portion of the significant perspectives about AI methods are:* First, a portion of the ML strategies, while being compelling, are computationally concentrated and require long time for learning tool. Consequently, to work with restricted capacity and computational assets of sensor organizations, it is a provoking errand to choose a pragmatic learning strategy which is successful in an assortment of situations. Second, under certain conditions it's anything but conceivable to get exact or exact information because of commotion. The last perspective is, now and again model examples or known marks possibly accessible for directing the learning cycle, while in other case it may not be available.

A Wireless Markup Language (WML) framework is based on distributed pattern recognition algorithms for leakage detection and size estimation, which is robust under different real-time scenarios and gathers the application specific real-time inputs [54]. Among different approaches for leakage detection, negative pressure wave technique is one of the effective tools to identify abrupt leakages in pipelines. The basic idea of these methods is that, when a leak occurs, a transient wave is generated which is reflected back to the measuring sensor node and can be sensed using pressure transducers [55].

The mix of AI and remote sensor network is a rich space of climate observing applications. Remote sensor organization can use the current AI calculations to give vigorous, dependable and independent observing.AI-based detections systems can be introduced for oil and gas pipelines circulation organizations [53]. The framework is grown natively and gives ability of detailing pipeline health, structure and condition related measurements extended over huge geographic regions [42]. The WML framework is conveyed effectively and gives exact constant observing of pipelines extended over enormous topographical areas. The normal recognition execution in perceiving spillages is 94.5% for support vector machines, 82.25% for Gaussian mixture model and 81.25% for Naive Bayes. Utilizing enormous informational indexes or a blend of these classifiers, the exhibition can be worked on further [39].

## REFERENCES

[1]   T. O.Ayodele, "Introduction to machine learning", in *New Advances in Machine Learning*. InTech, 2010.

[2]   A. H. Duffy, "The 'what' and 'how' of learning in design", *IEEE Expert*, vol. 12, no. 3, pp. 71–76, 1997.

[3] P. Langley and H. A. Simon, "Applications of machine learning and rule induction", *Communications of the ACM*, vol. 38, no. 11, pp. 54–64, 1995.

[4] L. Paradis and Q. Han, "A survey of fault management in wireless sensor networks", *Journal of Network and Systems Management*, vol. 15, no. 2, pp. 171–190, 2007.

[5] B. Krishnamachari, D. Estrin, and S. Wicker, "The impact of data aggregation in wireless sensor networks", in *22nd International Conference on Distributed Computing Systems Workshops*, 2002, pp. 575–578.

[6] J. Al-Karaki and A. Kamal, "Routing techniques in wireless sensor networks: A survey", *IEEE Wireless Communications*, vol. 11, no. 6, pp. 6–28, 2004.

[7] K. Romer and F. Mattern, "The design space of wireless sensornetworks", *IEEE Wireless Communications*, vol. 11, no. 6, pp. 54–61,2004.

[8] J. Wan, M. Chen, F. Xia, L. Di, and K. Zhou, "From machine-to-machine communications towards cyber-physical systems", *Computer Science and Information Systems*, vol. 10, pp. 1105–1128, 2013.

[9] Y. Bengio, "Learning deep architectures for AI", *Foundations and Trends in Machine Learning*, vol. 2, no. 1, pp. 1–127,2009.

[10] A. G. Hoffmann, "General limitations on machine learning *European Conference on Artificial Intelligence* 1990, pp. 345–347.

[11] M. Di and E. M. Joo, "A survey of machine learning in wireless sensor netoworks from networking and application perspectives", in *6th International Conference on Information, Communications Signal Processing*, 2007, pp. 1–5.

[12] A. Forster, "Machine learning techniques applied to wireless ad-hoc networks: Guide and survey", in *3rd International Conference on Intelligent Sensors, Sensor Networks and Information*. IEEE, 2007, pp. 365–370.

[13] M. Förster and L, Amy, *Machine Learning across the WSN Layers*. InTech, 2011.

[14] Y. Zhang, N. Meratnia, and P. Havinga, "Outlier detection techniques for wireless sensor networks: A survey", *IEEE Communications Surveys & Tutorials*, vol. 12, no. 2, pp. 159–170, 2010.

[15] V. J. Hodge and J. Austin, "A survey of outlier detection methodologies", *Artificial Intelligence Review*, vol. 22, no. 2, pp. 85–126, 2004.

[16] R. Kulkarni, A. Förster, and G. Venayagamoorthy, "Computational intelligence in wireless sensor networks: A survey", *IEEE Communications Surveys & Tutorials*, vol. 13, no. 1, pp. 68–96, 2011.

[17] S. Das, A. Abraham, and B. K. Panigrahi, *Computational Intelligence:Foundations, Perspectives, and Recent Trends*. John Wiley, 2010, pp. 1–37.

[18] Y. S. Abu-Mostafa, M. Magdon-Ismail, and H.-T. Lin, *Learning from Data*. AMLBook, 2012.

[19] O. Chapelle, B. Schlkopf, and A. Zien, *Semi-supervised Learning*. MIT Press, 2006, vol. 2.

[20] S. Kulkarni, G. Lugosi, and S. Venkatesh, "Learning pattern classification: A survey", *IEEE Transactions on Information Theory*, vol. 44, no. 6, pp. 2178–2206, 1998.

[21] M. Morelande, B. Moran, and M. Brazil, "Bayesian node localization in wireless sensor networks", in *IEEE International Conference on Acoustics, Speech and Signal Processing*, 2008, pp. 2545–2548.

[22] C.-H. Lu and L.-C. Fu, "Robust location-aware activity recognition using wireless sensor network in an attentive home", *IEEE Transactionson Automation Science and Engineering*, vol. 6, no. 4, pp. 598–609, 2009.

[23] A. Shareef, Y. Zhu, and M. Musavi, "Localization using neural networks in wireless sensor networks", in *Proceedings of the 1st International Conference on Mobile Wireless Middleware, Operating Systems, and Applications*, 2008, pp. 1–7.

[24] J. Winter, Y. Xu, and W.-C. Lee, "Energy efficient processing of nearest neighbor queries in location-aware sensor networks", in *2nd International Conference on Mobile and Ubiquitous Systems: Networking and Services*. IEEE, 2005, pp. 281–292.

[25] P. P. Jayaraman, A. Zaslavsky, and J. Delsing, "Intelligent processing of nearest neighbors queries using mobile data collectors in a location aware 3D wireless sensor network", in *Trends in Applied Intelligent Systems*. Springer, 2010, pp. 260–270.

[26] L. Yu, N. Wang, and X. Meng, "Real-time forest fire detection withwireless sensor networks", in *International Conference on Wireless Communications, Networking and Mobile Computing*, vol. 2, 2005, pp. 1214–1217.

[27] M. Bahrepour, N. Meratnia, M. Poel, Z. Taghikhaki, and P. J. Havinga, "Distributed event detection in wireless sensor networks for disaster management", in *2nd International Conference on Intelligent Networking and Collaborative Systems*. IEEE, 2010, pp. 507–512.

[28] M. Kim and M.-G. Park, "Bayesian statistical modeling of systemenergy saving effectiveness for MAC protocols of wireless sensornetworks", in *Software Engineering, Artificial Intelligence, Networking and Parallel/ Distributed Computing, ser. Studies in ComputationalIntelligence*. Springer, 2009, vol. 209, pp. 233–245.

[29] Y.-J. Shen and M.-S. Wang, "Broadcast scheduling in wireless sensor networks using fuzzy hopfield neural network", *Expert Systems with Applications*, vol. 34, no. 2, pp. 900–907,2008.

[30] R. V. Kulkarni and G. K. Venayagamoorthy, "Neural network based secure media access control protocol for wireless sensor networks", in *Proceedings of the 2009 International Joint Conference on Neural Networks, ser. IJCNN'09*. IEEE Press, 2009, pp. 3437–3444.

[31] D. Janakiram, V. AdiMallikarjuna Reddy, and A. Phani Kumar, "Outlier detection in wireless sensor networks using Bayesian belief networks", in *1st International Conference on Communication System Software and Middleware*. IEEE, 2006, pp. 1–6.

[32] J. W. Branch, C. Giannella, B. Szymanski, R. Wolff, and H. Kargupta, "In-network outlier detection in wireless sensor networks", *Knowledge and Information Systems*, vol. 34, no. 1, pp. 23–54, 2013.

[33] S. Kaplantzis, A. Shilton, N. Mani, and Y. Sekercioglu, "Detecting selective forwarding attacks in wireless sensor networks using support vector machines", in *3rd International Conference on Intelligent Sensors, Sensor Networks and Information*. IEEE, 2007, pp. 335–340.

[34] S. Rajasegarar, C. Leckie, M. Palaniswami, and J. Bezdek, "Quarter sphere based distributed anomaly detection in wireless sensor networks", in *International Conference on Communications*, 2007, pp. 3864–3869.

[35] A. Snow, P. Rastogi, and G. Weckman, "Assessing dependability of wireless networks using neural networks", in *Military Communications Conference*. IEEE, 2005, vol. 5, pp. 2809–2815.

[36] A. I. Moustapha and R. Selmic, "Wireless sensor network modeling using modified recurrent neural networks: Application to fault detection", *IEEE Transactions on Instrumentation and Measurement*, vol. 57, no. 5, pp. 981–988, 2008.

[37] Y. Wang, M. Martonosi, and L.-S. Peh, "Predicting link quality using supervised learning in wireless sensor networks", *ACM SIGMOBILE Mobile Computing and Communications Review*, vol. 11, no. 3, pp.71–83, 2007.

[38] K. Beyer, J. Goldstein, R. Ramakrishnan, and U. Shaft, "When is 'nearest neighbor' meaningful?" in *Database Theory*. Springer, 1999, pp. 217–235.

[39] T. O. Ayodele, "Types of machine learning algorithms", in New *Advances in Machine Learning*. InTech, 2010.

[40] S. R. Safavian and D. Landgrebe, "A survey of decision tree classifier methodology", *IEEE Transactions on Systems, Man and Cybernetics*, vol. 21, no. 3, pp. 660–674, 1991.

[41] R. Lippmann, "An introduction to computing with neural nets", *ASSPMagazine, IEEE*, vol. 4, no. 2, pp. 4–22, 1987.

[42] W. Dargie and C. Poellabauer, Localization. John Wiley, 2010, pp. 249–266.

[43] T. Kohonen, Self-organizing Maps, ser. Springer Series in Information Sciences. Springer, 2001, vol. 30.

[44] G. E. Hinton and R. R. Salakhutdinov, "Reducing the dimensionality of data with neural networks", *Science*, vol. 313, no. 5786, pp. 504–507, 2006.

[45] A. Christmann and I. Steinwart, *Support Vector Machines*. Springer, 2008.

[46] Z. Yang, N. Meratnia, and P. Havinga, "An online outlier detection technique for wireless sensor networks using unsupervised quarter-sphere

support vector machine", in *International Conference on Intelligent Sensors, Sensor Networks and Information Processing*. IEEE, 2008, pp. 151–156.

[47] Y. Chen, Y. Qin, Y. Xiang, J. Zhong, and X. Jiao, "Intrusion detection system based on immune algorithm and support vector machine in wireless sensor network", in *Information and Automation*, ser. Communications in Computer and Information Science. Springer, 2011, vol. 86, pp. 372–376.

[48] Y. Zhang, N. Meratnia, and P. J. Havinga, "Distributed online outlier detection in wireless sensor networks using ellipsoidal support vectormachine", *Ad Hoc Networks*, vol. 11, no. 3, pp. 1062–1074, 2013.

[49] W. Kim, J. Park, and H. Kim, "Target localization using ensemble support vector regression in wireless sensor networks", in *Wireless Communications and Networking Conference*, 2010, pp. 1–5.

[50] D. Tran and T. Nguyen, "Localization in wireless sensor networks based on support vector machines", *IEEE Transactions on Parallel and Distributed Systems*, vol. 19, no. 7, pp. 981–994, 2008.

[51] Yang, J. Yang, J. Xu, and D. Yang, "Area localization algorithm for mobile nodes in wireless sensor networks based on support vector machines", in *Mobile Ad-Hoc and Sensor Networks*. Springer, 2007, pp. 561–571.

[52] G. E. Box and G. C. Tiao, *Bayesian Inference in Statistical Analysis*. John Wiley, 2011, vol. 40.

[53] E. Rasmussen, "Gaussian processes for machine learning", in: *Adaptive Computation and Machine Learning*. Citeseer, 2006.

[54] S. Lee and T. Chung, "Data aggregation for wireless sensor networks using self-organizing map", in *Artificial Intelligence and Simulation*, ser. Lecture Notes in Computer Science. Springer, 2005, vol. 3397, pp. 508–517.

[55] R. Masiero, G. Quer, D. Munaretto, M. Rossi, J. Widmer, and M.Zorzi, "Data acquisition through joint compressive sensing and principal component analysis", in *Global Telecommunications Conference*. IEEE, 2009, pp. 1–6.

[56] R. Masiero, G. Quer, M. Rossi, and M. Zorzi, "A Bayesian analysis of compressive sensing data recovery in wireless sensor networks", in *International Conference on Ultra Modern Telecommunications Workshops*, 2009, pp. 1–6.

[57] A. Rooshenas, H. Rabiee, A. Movaghar, and M. Naderi, "Reducing the data transmission in wireless sensor networks using the principal component analysis", in *6th International Conference on Intelligent Sensors, Sensor Networks and Information Processing*. IEEE, 2010, pp. 133–138.

[58] S. Macua, P. Belanovic, and S. Zazo, "Consensus-based distributed principal component analysis in wireless sensor networks", in *11thInternational Workshop on Signal Processing Advances in Wireless Communications*, 2010, pp. 1–5.

[59] Y.-C. Tseng, Y.-C. Wang, K.-Y. Cheng, and Y.-Y. Hsieh, "iMouse: An integrated mobile surveillance and wireless sensor system", *Computer*, vol. 40, no. 6, pp. 60–66, 2007.

[60] Li, K. Wong, Y. H.Hu, and A. Sayeed, "Detection, classification, and tracking of targets", *IEEE Signal Processing Magazine*, vol. 19,no. 2, pp. 17–29, 2002.

[61] T. Kanungo, D. M. Mount, N. S. Netanyahu, C. D. Piatko, R. Silverman, and A. Y. Wu, "An efficient k-means clustering algorithm: Analysis and implementation", *IEEE Transactions on Pattern Analysis and Machine Intelligence*, vol. 24, no. 7, pp. 881–892, 2002.

[62] T. Jolliffe, *Principal Component Analysis*. Springer Verlag, 2002.

[63] Feldman, M. Schmidt, C. Sohler, D. Feldman, M. Schmidt, and C. Sohler, "Turning big data into tiny data: Constant-size coresets fork-means, PCA and projective clustering", in SODA, 2013, pp. 1434–1453.

[64] Watkins and P. Dayan, "Q-learning", *Machine Learning*, vol. 8, no. 3–4, pp. 279–292, 1992.

[65] R. Sun, S. Tatsumi, and G. Zhao, "Q-MAP: A novel multicast routing method in wireless ad hoc networks with multiagent reinforcement learning", in *Region 10 Conference on Computers, Communications, Control and Power Engineering*, vol. 1, 2002, pp. 667–670 vol.1.

[66] S. Dong, P. Agrawal, and K. Sivalingam, "Reinforcement learning based geographic routing protocol for UWB wireless sensor network", in *Global Telecommunications Conference*. IEEE, 2007, pp. 652–656.

[67] A. Förster and A. Murphy, "FROMS: Feedback routing for optimizing multiple sinks in WSN with reinforcement learning", in *3rd International Conference on Intelligent Sensors, Sensor Networks and Information*. IEEE, 2007, pp. 371–376.

[68] R. Arroyo-Valles, R. Alaiz-Rodriguez, A. Guerrero-Curieses, and J.Cid-Sueiro, "Q-probabilistic routing in wireless sensor networks", in *3rd International Conference on Intelligent Sensors, Sensor Networks and Information*. IEEE, 2007, pp. 1–6.

[69] Guestrin, P. Bodik, R. Thibaux, M. Paskin, and S. Madden, "Distributed regression: An efficient framework for modeling sensor network data", in *3rd International Symposium on Information Processing in Sensor Networks*, 2004, pp. 1–10.

[70] Barbancho, C. León, F. Molina, and A. Barbancho, "A new QoSrouting algorithm based on self-organizing maps for wireless sensor networks", *Telecommunication Systems*, vol. 36, pp. 73–83, 2007.

[71] B. Scholkopfand A. J. Smola, *Learning with Kernels: Support Vectormachines, Regularization, Optimization, and Beyond*. MIT Press, 2001.

[72] Kivinen, A. Smola, and R. Williamson, "Online learning with kernels", *IEEE Transactions on Signal Processing*, vol. 52, no. 8, pp. 2165–2176, 2004.

[73] Aiello and G. Rogerson, "Ultra-wideband wireless systems", *IEEE Microwave Magazine*, vol. 4, no. 2, pp. 36–47, 2003.

[74] R. Rajagopalan and P. Varshney, "Data-aggregation techniques in sensor networks: A survey", *IEEE Communications Surveys & Tutorials*, vol. 8, no. 4, pp. 48–63, 2006.

[75] G. V. Crosby, N. Pissinou, and J. Gadze, "A framework for trust-based cluster head election in wireless sensor networks", in *2nd IEEE Workshop on Dependability and Security in Sensor Networks and Systems*, 2006, pp. 10–22.

[76] J.-M. Kim, S.-H. Park, Y.-J. Han, and T.-M. Chung, "CHEF: Clusterhead election mechanism using fuzzy logic in wireless sensor networks", in *10th International Conference on Advanced Communication Technology*, vol. 1. IEEE, 2008, pp. 654–659.

[77] S. Soro and W. Heinzelman, "Prolonging the lifetime of wireless sensor networks via unequal clustering", in *19th IEEE International Parallel and Distributed Processing Symposium*, 2005, pp. 4–8.

[78] A. A. Abbasi and M. Younis, "A survey on clustering algorithms for wireless sensor networks",*Computer Communications*, vol. 30, no. 14, pp. 2826–2841,2007.

[79] H. He, Z. Zhu, and E. Makinen, "A neural network model to minimize the connected dominating set for self-configuration of wireless sensor networks", *IEEE Transactions on Neural Networks*, vol. 20, no. 6, pp. 973–982, 2009.

[80] G. Ahmed, N. M. Khan, Z. Khalid, and R. Ramer, "Cluster head selection using decision trees for wireless sensor networks", in *International Conference on Intelligent Sensors, Sensor Networks and Information Processing*. IEEE, 2008, pp. 173–178.

[81] Ertin, "Gaussian process models for censored sensor readings", in *14th Workshop on Statistical Signal Processing*. IEEE, 2007, pp.665–669.

[82] J. Kho, A. Rogers, and N. R. Jennings, "Decentralized control of adaptive sampling in wireless sensor networks", *ACM Transactions on Sensor Networks (TOSN)*, vol. 5, no. 3, pp. 19:1–19:35, 2009.

# CHAPTER 5

# Applications of Machine Learning – Fire Detection

## R. M. Mehra[1] and Rashmi Priyadarshini[2]

*Sharda University, Greater Noida, Uttar Pradesh, India*
*Email ID: rm.mehra@sharda.ac.in; rashmi.priyadarshini@sharda.ac.in*

## 5.1 INTRODUCTION

Early identification of a fire and the sounding of an appropriate alarm have proven to be important factors in averting major fire losses in the past. Fire detection and alarm systems that are properly built and maintained can help to enhance tenant and emergency responder survival while lowering property losses. Fire detection and alarm systems, together with automated fire suppression systems, are part of the active fire protection systems present in many occupancies. Adopted buildings and/or fire codes may mandate the installation of fire detection and alarm systems to this purpose. These systems almost always need the installation and maintenance of skilled professionals. Today hundreds of thousands of lives are saved thanks to a wide range of new technology. Preventive technologies work to stop a fire from starting in the first place. Smoke, heat, and flame detectors are examples of these devices, which can assist offer early warning of fires. Suppressive technologies aid in the extinguishment of a raging fire. To assist prevent fires from spreading, a variety of sprinkler systems are now available. Fire alarm systems, on the other hand, aren't exactly cutting-edge in terms of societal progress; creative firms are devising novel approaches to fire and gas-related risks to get the most out of the technology and guarantee its long-term viability. To ensure the highest level of safety, we must constantly assess not just current technology and its use, but also existing legislation, rules, and enforcement, as well as making changes in these areas.

Poor visibility is a common issue for firefighters while attempting to rescue people trapped in a building engulfed in flames and smoke. It becomes difficult to detect anything on their route of rescue, such as a door, a staircase, or any other impediment, thus causing a delay in the rescue operation. Such difficulties are alarming, especially when a person's life is on the line. This chapter covers the fundamentals of machine learning and how it may be used to identify fires. It will also explore how tools like computer vision, machine learning, and deep learning may aid front-line employees in making better decisions in stressful situations. The information provided by sensors would be beneficial in terms of rescue time and the lives of both the victim and the firefighter. The promise of a more out-of-the-box technical innovation will result from the creative revolution in fire alarm systems, as well as artificial intelligence, connected systems, and the smart city project [1].

### 5.1.1 Artificial Intelligence (AI)

The current buzzword sweeping the global corporate scene is machine learning. It's snatched the public's attention, evoking ideas of self-learning AI and robotics in the future. Machine learning has opened the way for technical advancements and tools in business that would have been unthinkable only a few years ago. It fuels the revolutionary inventions that enable our modern lifestyles, from prediction engines to internet TV live streaming.

The phrase "machine learning" refers to a collection of techniques and technologies that enable computers to learn and adapt on their own. AI can learn without being explicitly taught to do the required action thanks to machine learning techniques. The machine learning algorithm learns a pattern from sample inputs and predicts and performs tasks entirely based on the learned pattern. The machine learning algorithm anticipates and performs tasks purely based on the learned pattern, rather than a predetermined program command, using sample inputs. Machine learning comes to the rescue in a variety of situations when rigid methods aren't feasible. It will learn the new procedure from past patterns and put what it has learned into action [2].

The way our email providers assist us deal with spam is one of the machine learning applications we are acquainted with. Spam filters employ an algorithm to detect and route new types of junk mail to your spam box. Machine learning algorithms are being used by some e-commerce firms in conjunction with other IT security solutions to avoid fraud and enhance their performance.

The following are the top real-world machine learning applications for 2021 [3]:

Features of Social Media
Recommendations for Products
Recognition of Images
Analysis of Public Opinion
Automating Access Control for Employees
Conservation of Marine Wildlife
Healthcare Efficiency and Medical Services Regulation
Banking Domain Language Translation to Predict Potential Heart Failure

## 5.1.2 The Most Important Trends in Fire Alarm Systems

In comparison to other high-demand smart gadgets, fire alarm system technology progress has been sluggish. Manufacturers throughout the world concentrate their investing in the development of high-return products and to customers who are linked, in particular, to quickly shifting lifestyle patterns. Creative businesses are coming up with novel ways to deal with fire and gas-related dangers. Smoke, heat, and flame detectors are examples of these devices, which can offer early warning of fires. Suppressive technologies aid in the extinguishment of a raging fire. Today, we may choose from a variety of sprinkler systems to assist in preventing the spread of flames.

To get the most out of technology while maintaining the highest level of safety, we should examine current technology and its application, as well as rules, which will lead to even more innovative technological solutions. Some of the most noteworthy market trends are shown in Figure 5.1 [4].

Aspiration sensors can detect tiny smoke particles in the air, with precise measurements; there are two types of aspiration technology available on the market. The system draws air samples to check for potential dangers. A laser-based technique will grow the market of sensors. We may see a significant growth in demand for this technology in the near future.

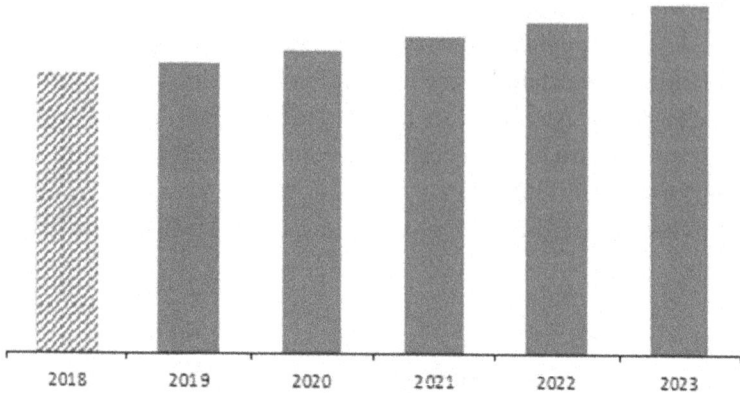

FIGURE 5.1    Worldwide fire alarm system in market.

## 5.1.3 Internet of Things (IoT) in Fire Safety Systems

One of the many areas where the Internet of Things (IoT) may provide remarkable benefits is fire protection. IoT-enabled fire detectors have already been released by a number of major fire safety businesses.

## 5.1.4 Connected Detectors with IoT Capability

In the marketplace, there are a number of well-known firms. Smoke and carbon monoxide detectors that are linked for home usage are available. These connected devices can alert you to a fire or carbon monoxide leak and communicate with the thermostat to turn off the heater. Mobile apps allow users to access the detectors from anywhere. The detectors sound a local alarm and transmit alerts to the phone in the case of an alarm. Some of the most well-known smart detectors available today are Nest and Ring.

Existing detectors can also be retrofitted with IoT technology to enable connectivity. Kidde, a major fire safety device maker, provides a monitor that can be used to link existing fire detectors. Users do not need to replace all detectors while using a monitor. When the monitor detects the smoke, it sends an alarm to its app based on the frequency of smoke and CO detectors. Multiple detectors on the floor can be covered by a single monitor.

Roost is a firm that provides consumers with another retrofit alternative. Users may add smart features to their old detectors by replacing their standard batteries with Roost batteries. When the detector sounds, the battery might send a push notification. The software also allows users to check the remaining battery life at any point of time. Connectivity as well as a number of other smart detectors are available. Goods are provided by large businesses like First Alert and new startups like Leeo [5].

## 5.2 TECHNOLOGICAL ADVANCES IN CENTRAL ALARM SYSTEMS

The detectors of a central alarm system are all connected to a central controller and relay their signals to it immediately. When the controller receives alarm input from the detecting sensors, it activates warning devices such as horns, strobe lights, and speakers to alert the inhabitants. Central alarm and evacuation systems are becoming increasingly connected and intelligent. Mass notification systems now include a paging component that transmits live audio instructions across the site in the case of an emergency. Most mass notification systems now include a wide range of emergency messages, including severe weather warnings, security alerts, and amber alerts, among others. One of the most advanced features of mass notification systems is the ability to send SMS, text, email, popup, app message, and push notifications to specified recipients, enabling for fast and effective awareness. Because of the IoT and other current technology, progress is conceivable. It can also be used in emergency situations. Sensor and detector data, as well as other monitoring data, may be used with algorithms and analytics to aid in the development of better emergency or evacuation plans. To improve evacuation preparations, analytics may examine a variety of factors, including the number of people in the building, building maps, the location of the fire, and the rate at which the fire spreads, as well as the fire's direction. Analytics-based evacuations plans can assist minimize crowding by guiding individuals from various parts of the building to the most effective evacuation path.

### 5.2.1 Detection using Multiple Sensors

Wireless fire alarm systems will experience a rise in demand as they become less expensive and more practical [6].

### 5.2.2 Voice Detection Systems

Systems that use a pre-recorded message to warn of danger are gaining momentum and show no signs of slowing down in the foreseeable future. The voice alerts system makes possible the delivery of particular life safety directives and evacuation safety procedures. This technique is useful not just in fires, but also in other types of situations, for instance as a result of a dangerous chemical spill, an intruder, or inclement weather. Furthermore, these systems allow building managers to communicate with residents through voice alarm system to provide basic evacuation instructions; studies have shown that people react more positively to spoken instructions than to simple sirens and recorded speech.

Remote monitoring helps to reduce risks. Modern fire alarm systems may be configured to deliver alerts both on-site and off-site. All regulatory requirements and legislation must be met by the fire alarm system, including the central remote monitoring station. Even if the capacity to monitor systems remotely raises the cost of the system, as well as the cost of maintenance and operations, it is well worth it. The advantages of such features are enormous. Multiple buildings can be linked to a single central monitoring station, allowing facility managers to better monitor alarm systems, and to maintain schedule testing and maintenance procedures, as well as keep track of records and reports. As a result, creating organizations presents both problems and possibilities. Making sense of big, complicated data sets is one of the problems. One method is to use predictive analytics and machine learning techniques. The prospective return on investment is observable, immediate, and real. The IoT will supply more data from a larger range of sources, and at a quicker rate. To deal with the growing complexity of data and adapt to it events in a timely way, the analytics value chain will need to be much more automated and frictionless. Machine learning-based maintenance apps are a potential way to make the most of IoT data. Even with the most advanced technologies, we will never be able to completely avoid accidents.

Although safety technology has advanced significantly, it is still insufficient and will likely never be perfect. Disasters frequently occur as a result of poor execution and planning and not because of broken equipment or a lack of adequate technology. It won't assist in improving safety unless the cutting-edge safety technologies are deployed and maintained correctly. Regulations, processes, and regulations must all be in place and strictly adhered to. Think about what worked and what didn't in your business in the last few years, as well as what you may do to improve. Prepare to make some changes and grow your business by learning how to improve the engagement, efficiency, and effectiveness of your fire alarm system [7].

## 5.3 AI APPLICATIONS IN FIRE AND SAFETY

This chapter looks at how computer vision, for example, may help in fire safety. In a stressful scenario, machine learning and deep learning can assist front-line personnel in making better informed decisions. The information provided by the sensors would be beneficial in terms of rescue time and the lives of both the victim and the firefighter.

### 5.3.1 AI for Front-Line Personnel

As mentioned, poor visibility is a typical issue for firefighters while attempting to rescue people trapped in a building engulfed in flames and smoke. To address this problem, the technique is based on rudimentary computer vision. Data from the camera is utilized to enhance the eyesight of fire authorities by placing a camera on top of the helmet. The following is the flow of the whole procedure. The camera's picture frames are transmitted to a processor, which performs simple matrix operations. The processor will use various filters, such as a Sobel operator, to do edge detection and contour detection. The processor's result will be shown on the fire official's AR glasses. The AR glasses' vision would be similar to the illustration above. The fire officials may effectively detect the surroundings using this basic technique. This makes things simple because the firefighter can quickly detect things in their environment, speeding up the entire rescue procedure.

For fire authorities, an AI and IoT-based solution is available. The heat from the flames can induce suffocation, convulsions, and even heart attacks, in addition to visual issues – so much so that one of the aforementioned dangers is responsible for the majority of fire official deaths. An IoT-based solution can be implemented to reduce such risks. Sensors such as oxygen level detectors are used to see if the firefighter's oxygen level is within normal limits. ECG sensors can determine the level of physiological arousal a person is experiencing in order to make better inferences about their psychological state [8]. The AI and IoT-based solution adopted by the fire authorities is shown in Figure 5.2.

Body temperature sensors track changes in the human body's temperature to ensure that the temperature range stays within acceptable limits. A recommendation system with a large amount of data alerts the official if there is an abnormality in the data. A machine learning model may be used to categorize data from numerous sensors in the range into three categories: safe, dangerous, and urgent. Many techniques, such as decision

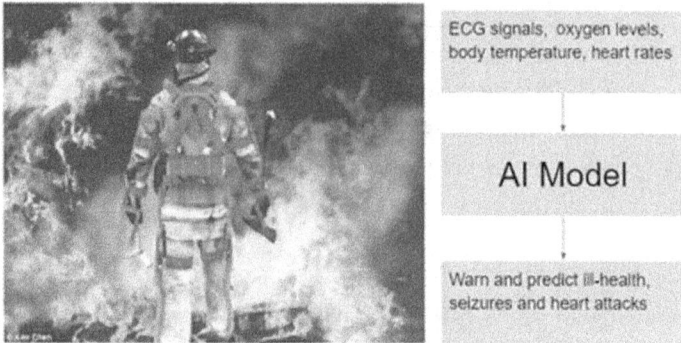

FIGURE 5.2    AI and IoT-based solution used by fire authorities.

trees, random forests, ANNs, and K-means clustering, can be utilized for this purpose. The data flow is as follows. The processor receives data from the sensors. A probabilistic or categorical output is generated by the machine learning model. The user will be able to use the output data utilized by the supporting crew to communicate with the officials over the phone to alert them of danger.

AUDREY (Assistant for Data Understanding through Reasoning, Extraction, and Synthesis) is a NASA-developed AI solution. We covered how certain sensors can ease a stressful situation by interpreting data from the respondent's body in the preceding section. AUDREY, on the other hand, examines the surrounding facts and suggests the safest course of action. If AUDREY detects a high level of hazardous gases such as carbon monoxide, it can alert the respondent and advise them to be extra cautious. AUDREY can gather data about temperature, gases, and other danger signs using artificial intelligence in order to safely direct a team of first responders through the flames.

## 5.3.2  AI to Combat Wildfires

Image-based and sensor-based methods are utilized [9] to battle forest fires using AI. A convolutional neural network is trained to detect fires via an image-based method. Preparing the dataset, annotating the dataset, training the model, and validating the model are all steps in the process. The deep learning model is typically employed on drones that are used to identify the existence of wildfires for surveillance reasons. The task is quite similar to that of a model that detects objects. After the model has learned the flaw's characteristics, it is possible to utilize the flames from the data and annotations for detecting purposes. Mobile Net is the most often used model because it combines high accuracy with low computing cost because to the use of depth-wise convolution. Using the k-nearest neighbor method, this may be expanded to detect the severity of the occurrence. This strategy is related to the YOLO (You Only Live Once) paradigm. The drones detect the fires and alert the authorities, who then take appropriate action. Deep learning is applied in this way to lessen the severity of the disaster [10].

## 5.4 SENSOR-BASED STRATEGY

A sensor-based method uses a number of sensors scattered across a forest to create a cumulative prediction. Forest fires are a common occurrence. Sensors are used to monitor carbon dioxide, hydrogen sulphide, carbon monoxide, and oxygen in the environment. This information is used to identify wildfires, as well as changes in the surrounding temperature and humidity. This approach uses a machine learning model that has been trained on millions of datasets and is capable of producing accurate predictions. When the machine learning model identifies an anomaly, it sends out an alert warning to the forest service. Classification models like the random forest may also be used to identify forest fires. Predicting the occurrence of an event is a more advanced use of the same concept. This is a time series problem that deep learning models like LSTM or recurrent neural networks can solve. As a result, AI still has plenty of opportunity to grow in the future. Fire and safety are two fields where I work. These are only a few examples of applications that might be utilized in the case of a disaster to protect the environment and people's lives [11].

### 5.4.1 System Classifications of the NFPA

An auxiliary fire alarm system is a system that is used in addition to the main fire alarm system – a system for delivering a fire alarm to the public fire service communication center that is connected to a municipal fire alarm system. Auxiliary fire alarm system alarms are received at the public fire service communication center on the same equipment and using the same methods as alarms transmitted manually from municipal fire alarm boxes located on streets.

### 5.4.2 Fire Alarm System in the Central Station

A system or a collection of systems in which circuit and device activities are automatically transmitted to be recorded, maintained, and monitored from a central station with component and expert servers and operators who, upon receiving a signal, take the action required by this code. A service like this has to be monitored and maintained – by an individual, a company, or a group of people, a system or a group of systems in which circuit and device activities are automatically transmitted to, recorded in, maintained by and monitored from a central station with component and expert servers and operators who, upon receiving a signal, take the action required by this code. It is necessary to monitor and administer such a service. A person, company, or organization that is in the business of making money is used to govern and run the service by providing, maintaining, or monitoring supervised weapon systems. Combination System. A fire alarm system that, in whole or in part, shares components with a non-fire signaling system. Household Fire Alarm System. A group of devices that use a fire alarm control (panel) to generate an alert signal in the home to notify residents of a fire. They need to know if there's a fire so they can get out.

### 5.4.3 Municipal Fire Alarm System

Alarm-initiating devices, receiving equipment, and connecting circuits (other than a public telephone network) are used to send alarms from street sites to the public fire department by the communications center.

### 5.4.4 Fire Alarm System with a Proprietary Supervising Station

A fire alarm system is installed on contiguous and noncontiguous properties under one owner's management from a proprietary supervising station located at the protected property, or at one of several noncontiguous protected properties where trained, competent workers are constantly present. Supervising station, power supply, signal starting devices, and initiating device are all private circuits; signal notification appliances, equipment for the automated, permanent visual recording of such signals, and equipment for triggering the activation of emergency building control services are all examples of proprietary circuits.

### 5.4.5 Protected Premises Fire Alarm System (Local)

Premises Fire Alarm System (Local): A sound-based security system for your business. Fire Alarm System for Premises (Local): A protected premises system that sounds an alarm at the protected premises asthe result of manual operation, a fire alarm box or the operation of protection equipmentor systems, such as water flowing in a sprinkler system, the discharge of carbon dioxide, the detection of smoke or the detection of heat. Alarms are sent from street sites to the communications center via alarm-initiating devices, receiving equipment, and connecting wires in this system.

### 5.4.6 Remote Supervising Station Fire Alarm System

A system configured to deliver alarm, supervisory, and problem signals from one or more protected sites to a remote site where appropriate action is taken in accordance with this code. The National Fire Alarm Code provides terminology that is often used when talking about fire detection and alarm systems. Alarm: A fire-danger notice Supervisory (Signal) sees Signal Trouble (Signal) sees Signal. An electrical or other way of communicating a status indicator. An alarm has been sounded. Evacuation warning is a signal that indicates an emergency that necessitates quick response, such as a fire signal [12]. A unique signal designed to alert the building's inhabitants that the building must be evacuated. Signal for a fire alarm. A fire alarm-initiating equipment, such as a manual fire alarm box, automatic fire detector, or water flows, sends out signal. A manual fire alarm box, automatic fire detector, waterflow switch, or other device in which activation is initiative of the presence of a fire or fire signature.

## 5.5 SUPERVISORY SIGNAL

A signal indicating the need for action in the supervision of guard tours, the fire suppression system or equipment, or aspects of related systems that needs to be repaired.

### 5.5.1 Trouble Signal and (Device/Circuit) Supervision

A signal generated by a fire alarm system or device that indicates a failure in a circuit or component that is being monitored. The control panel contains an auditory signal device that is activated in conjunction with the visual indication in the event of an alarm, problem, or supervisory condition.

Fire detection and alarm systems vary from other commercial items in that they are supposed to have a specific purpose, namely, to protect humans from the effects of fire. Fire detection and alarm systems must be checked on a regular basis to verify that all system components are in proper working order so that they can detect and report a fire as intended. This is referred to as supervision, and virtually everything in fire detection and alarm system is supervised to verify that it is working. From the power supply, to the notification appliances, to the beginning device circuits in a traditional system, to individual intelligent fire detectors and other intelligent devices, everything in an intelligent system must be managed. A fault in a supervised circuit indicates a problem on the control panel. The control panel provides them eans to operate the system for functions such as silencing and un-silencing (resounding) the audible notification appliances as well as resetting the system.

The control panel houses the system's power supply as well as the circuits and controls for the starting devices, which might be conventional or intelligent (or both), as well as the circuits and controls for the notification appliance. According to the system set-up and application, the control panel monitors and checks the initiating devices, as well as the notification devices and other outputs [13].

Optical fire sensors, also known as "volumetric" sensors, work in tandem with traditional point sensors, for instance people can use smoke and heat detectors to get early alerts about fires. Cameras using image processing software can detect flames quicker than point sensors and provide more information than their traditional equivalents in terms of size, growth, and direction. A system capable of detecting both smoke and flames is more likely to detect fire sooner than one that can just detect smoke [14].

This chapter looks at how, as an early warning system, an RGB camera is utilized to detect flame and smoke. Warning systems for fires that occur within a radius of one to 20 meters from the camera is the subject of this study [15]. QuickBlaze is the name we gave to our system.

## 5.6 RESEARCH METHODOLOGY

The QuickBlaze architecture in general is shown in Figure 5.3.

### 5.6.1 Color

We assume that each frame has a fair color distribution; we utilize the "gray-world" technique, which is one of the simplest illuminant estimation methods. For each frame, the average of the picture intensities in the R, G, and B planes is computed. The "gray-value" of a photograph is the amount of black and white in the image, resulting in a vector of three intensities. The gray value is then normalized to the frame's average intensity. Each R, G, and B plane is scaled independently using a multiplication factor in the R, G, and B planes. If a scaled value approaches the greatest intensity achievable, it is clamped to the maximum. The field of view of fire detection vision sensors is commonly believed to be fixed at a given position with

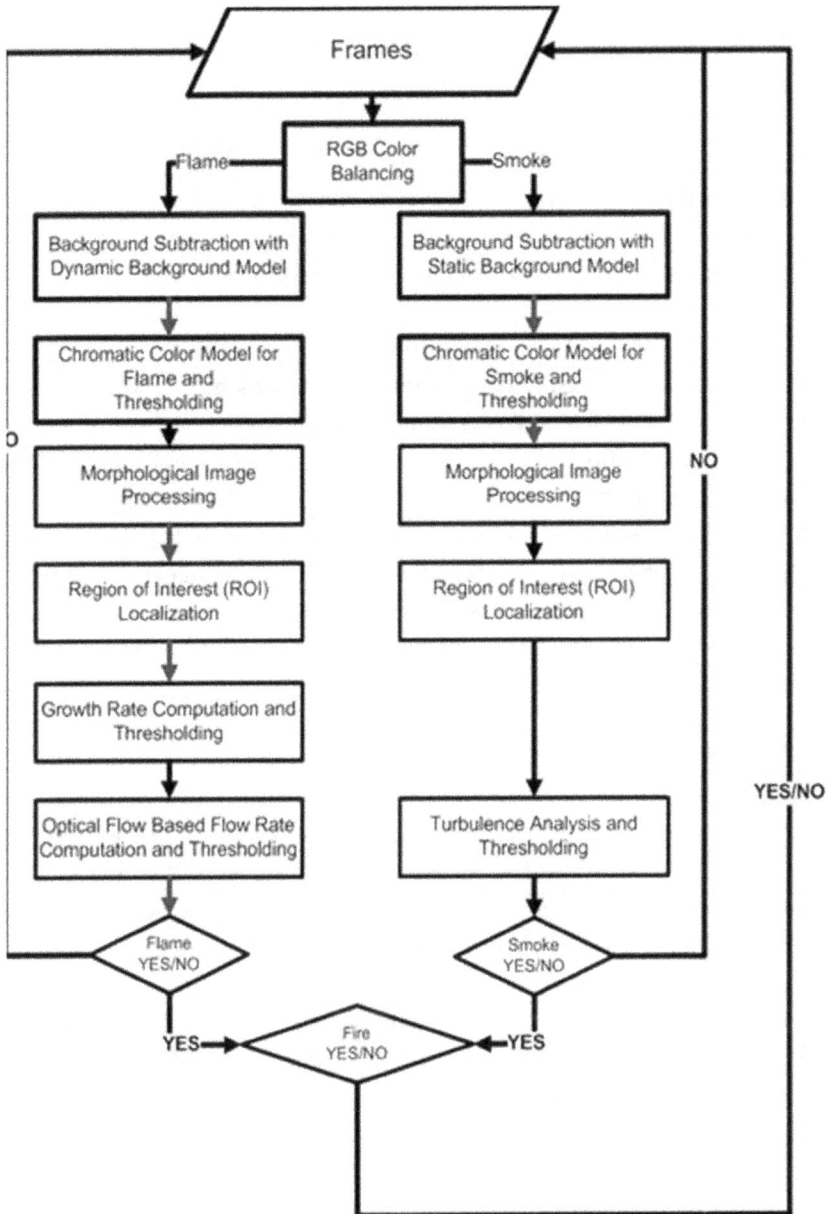

FIGURE 5.3 QuickBlaze architecture.

orientation and backdrop scene remaining constant. Using a background model, the foreground objects are extracted and motion areas are identified and segmented.

### 5.6.2 Chromatic Filtering

Flames are often red-yellow or reddish in hue, whereas the color of smoke varies from bluish white to white, at least at the commencement of a fire. To detect flame and smoke zones, we utilize several color models.

### 5.6.3 Image Morphology Processing

The resulting candidate areas are often noisy after simple filtering based on motion and color cues. We carry out morphology to reduce noise and select final candidate areas. Morphological image processing was done on both the flame and smoke pixel maps. To begin, we seal holes and eliminate minor noise components using basic approaches. Then, based on the length of the region's perimeter, we acquire linked components and eliminate tiny components. Finally, for each person who has made it thus far, through the potential fire and smoke region, we receive a chain code.

### 5.6.4 Candidate Regions' Geographical Location

Multiple potential flame and smoke areas may exist in a single frame. For growth rate and flow rate in flame regions, or turbulence in smoke regions, each candidate zone should be monitored and localized individually. Each zone may be categorized as freshly generated or previously existent when tracking regions from frame to frame. We do this by logically adding the masks for the current and prior frames (obtained after filtering for motion and color morphology). A foreground area is deemed freshly born if all of its values are zero. If a part of the values in a specific foreground region are non-zero, the region is linked to the previously recognized region. Before being regarded proof of a construct at, a candidate area must survive (be linked to a region in the previous frame) for a certain number of frames, real fire, both in terms of smoke and flame (see the next subsection for details).

## 5.7 DETECTION OF SMOKE

To complete the collection of observed areas for smoke, we utilize simple motion, color, and connection criteria.

## 5.7.1 Flame Detection Algorithm

When the behavior of false positive flame areas is inconsistent with flame physics, we may remove them using pixel color change and turbulence analysis, much as we do with false positive smoke regions. Fire is characterized by turbulent flames. Because a gas's density, plume will rise above the burning item, generating upward motion, inversely proportionate to its temperature, as seen in Figure 5.3. The burning item in the hot plume is forced to go higher as air rises above it. The buoyancy of the hot plume reduces as the surrounding air cools it, causing it to stop rising and start descending. Eddies happen when cold air is pushed into the fire plume, creating rising vortices and turbulence (a process called entrainment). To define the usual dynamics we use two heuristics to filter out locations that are unlikely to contain flame: analysis of growth rates and flow rates.

## 5.7.2 Experimentation Findings

Our tests are aimed at assessing the system's ability to detect flames as quickly as possible once they start burning, as well as crucial analytical indications. The reaction time, frame processing rate, and false error detection rate are used to evaluate the algorithm's performance. We manually modified the threshold values to get the system to work correctly. On one of three training videos, an orange balloon distracts the viewer. We picked the first option based on human observation of the event frame that contains a flame or smoke A selection of non-flame and non-smoke frames from the test videos are shown in Figure 5.4, which gives a selection of frames from test footage that includes flames or smoke.

We discovered that QuickBlaze outperforms the commercial system in terms of response time, speed, and fire localization. QuickBlaze may be utilized in any situation as long as the fire or smoke is visible to the video sensor, and it will likely respond faster than smoke detectors. This does suggest, however, that it may be most helpful in large open spaces (commercial). Commercial or industrial settings are more viable, but residential sites with several small rooms are less so. Additionally, a single sensor system may not be the most effective detector in all circumstances. Of course, classic point-based sensors may be used to supplement the video-based method described here in interior situations [16].

Early detection of fires before they evolve into catastrophic events is critical in preventing fire disasters and saving people's lives and property. Despite the fact that fire and smoke detection sensors are commonly used

FIGURE 5.4    The earliest frame showing flame or smoke, as determined by human observations, is presented in each example. Despite the fact that M-1, M-3, and M-4 are identical to the films in the training set, they were recorded at various times and with distinct target objects.

in interior spaces, they usually need the fire to burn for a certain amount of time. It takes a long time to create a big volume of smoke and then set off the alarm. Furthermore, these gadgets are not suitable for use in large-scale outdoor situations, such as forests or wild areas. Vision-based fire detection systems, on the other hand, record pictures from cameras and identify fire instantly, making them ideal for early fire detection. This type of device is inexpensive and simple to set up. We presented a vision-based fire detection system that may be used with a non-stationary camera in this chapter. For large-scale forest fire detection, our fire detection system may be installed to unmanned aerial vehicles.

In the literature, there have been various approaches for vision-based fire detection. Color model, movement, and spatial analogue. Fire has extremely distinct properties as compared to other objects, therefore kinematic, spatial, and temporal elements are largely employed. Almost all of the suggested approaches use a similar detection pipeline, in which moving pixels are first identified via background subtraction, and then a color model is used to identify fire color areas. The irregular and flickering features of fire are detected by analyzing these areas geographically and temporally. These approaches only work with fixed cameras, that is, in surveillance settings, because motion is the primary characteristic. Our approach is not constrained by such restrictions in our work. To learn feature representations from the data, we use a current, strong deep learning approach to train discriminative classifiers for fire area identification using data representations. We employ deep convolutional neural networks as the learning machine in particular [17].

### 5.7.3 Dataset

We assemble a fire detection dataset from a variety of internet sources, from which the majority of the video data was obtained, which is widely utilized in current fire detection literature. There are 25 films in the present collection. There are 21 positive (fire) sequences and 4negative (non-fire) sequences, all of which were captured in a woodland setting. To train an image-based fire classifier, we extract images from the videos at a sample rate of 5, or 1image per 5frames. These images have been resized to fit the usual dimensions. We used 323 to manually designate the existing fire patches' localization. In this work, we employ a tiny subset to build and test our algorithms as a proof of concept. Table 5.1 shows the statistics of the utilized train and test pictures, as well as the number of annotated patches. Figure 5.5 shows several sample photos with annotations. Depending on the level of acceptance, we may make this benchmark publicly available in order to assist future researchers in evaluating vision-based fire detectors [18].

TABLE 5.1   Statistics of the Proposed Fire Detection Dataset

| Set | Images | Patches | Positive Patches | Negative Patches |
|---|---|---|---|---|
| Train Set | 178 | 12460 | 1307 | 11153 |
| Test Set | 59 | 4130 | 539 | 3591 |

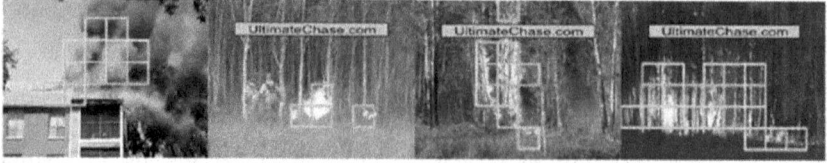

FIGURE 5.5   Sample images from the proposed benchmark and their patch-wise ground truth.

Modern structures are generally fitted with the newest materials to offer safety and comfort, thanks to the progress of science and technology. When it comes to safety, one of the most significant concerns that all architects want to address is security, which is one of their top considerations when creating a building. According to the National Fire Protection Association of the United States, in 2013, departments replied to an estimated 1,240,000 requests for information; 13,240 individuals were killed; 15,925 were injured; and $11.5 billion in property damage was projected as a result of these events. Significant major events resulted in a substantial number of fatalities [18].

Fires have had a destructive effect on both property and human lives for a very long time. As a result, humans have devised a variety of fire-fighting strategies to address the problem. The novel method forecasts the likelihood of a fire breaking out. However, predicting the incidence of fire remains a difficult challenge because fire is linked to a plethora of factors that influence its incidence, and the overall number of fire occurrences each day is quite low. This indicates that the data set for fire incidence is sparser, as the majority of the data sets' values are zero. As a result, in order to address this problem, assessing the risks of the prediction of the occurrence of fire has been implemented. Boosting, super vector machines, and logistic regression are just a few of the machine learning methods utilized in this technique. Fire detection has gotten more attention across the world as a result of the health and environmental effects of fires, as well as the fact that fire regimes have been changing. When compared to the existing models for predicting the incidence of fire, the accuracy of predicting using machine learning models is significantly higher. Furthermore, everyone anticipates that machine learning models will aid in the evaluation of the possibilities.

The severity of fire-related dangers in industries and structures will be assessed [19].

This is also an important study topic in the field of function approximation. Various data regression artificial neural network techniques, including Radial Basis Function and Multi-layer Perceptron, have been introduced.

## REFERENCES

1. M. Harman, The role of artificial intelligence in software engineering. In *Proceedings of the 2012 First International Workshop on Realizing AI Synergies in Software Engineering (RAISE)*,Zurich, Switzerland, 5 June 2012, pp. 1–6.

2. *Machine Learning Methods for Network Intrusion Detection and Intrusion Prevention Systems.* Zheni Svetoslavova Stefanova Graduate Theses and Dissertations Graduate School (Thesis) University of South Florida Scholar Commons.2018

3. Nutan Farah Haq, Abdur Rahman Onik, Md. Avishek Khan Hridoy, Musharrat Rafni, and Faisal Muhammad Shah', Application of Machine Learning Approaches in Intrusion Detection System: A Survey, Dewan Md. Farid, *International Journal of Advanced Research in Artificial Intelligence (IJARAI)*, Vol. 4, No. 3, 2015. pp. 9–18

4. L. Salhi, T. Silverston, T. Yamazaki and T. Miyoshi, Early Detection System for Gas Leakage and Fire in Smart Home Using Machine Learning, 2019 *IEEE International Conference on Consumer Electronics (ICCE)*, 2019, pp. 1–6.

5. Mohammad Sultan Mahmud, Md. Shohidul Islam and Md. Ashiqur Rahman, Smart Fire Detection System with Early Notifications Using Machine Learning, *International Journal of Computational Intelligence and Applications* Vol. 16, No. 2,2017pp 1–17

6. Y. Lim, S. Lim, J. Choi, S. Cho, C.K. Kim, Y.W. Lee, H. Zhang, H. Hu, B. Xu, J. Li et al. A Fire Detection and Rescue Support Framework with Wireless Sensor Networks. In *Proceedings of the 2007 International Conference on Convergence Information Technology (ICCIT 2007)*, Gyeongju, Korea, 21–23 November 2007, pp. 135–138.

7. A. Imteaj, T. Rahman, M. K. Hossain, M. S. Alam, S. A. Rahat, An IoT Based Fire Alarming and Authentication System for Workhouse using Raspberry Pi 3. In *Proceedings of the 2017 International Conference on Electrical, Computer and Communication Engineering (ECCE)*, Cox's Bazar, Bangladesh, 16–18 February 2017, pp. 899–904.

8. Jun Hong Park, Seunggi Lee, Seongjin Yun, Hanjin Kim and Won-Tae Kim, Dependable Fire Detection System with Multifunctional Artificial Intelligence Framework, *Computer Science Sensors* 2019, Vol.19, pp. 1–22

9. Wiame Benzekri, Ali El Moussati, Omar Moussaoui and Mohammed Berrajaa, Early Forest Fire Detection System using Wireless Sensor Network and Deep Learning, *(IJACSA) International Journal of Advanced Computer Science and Applications*, Vol. 11, No. 5, 2020, pp. 496–503

10. Faisal Saeed, Anand Paul, P. Karthigai kumar and Anand Nayyar *Convolutional Neural Network Based Early Fire Detection*, 2020, Vol. 8, no. 12, pp. 127–130

11. S. Pouyanfar, S. C. Chen, Semantic Event Detection using ensemble deep Learning. In *Proceedings of the 2016 IEEE International Symposium on Multimedia (ISM)*, San Jose, CA, USA, 11–13 December 2016, pp. 203–208.

12. M. J. Karter, *False Alarm Activity in the US 2012*, National Fire Protection Association: Quincy, MA, USA, 2013.

13. Fengju Bu, Intelligent and Vision-Based Fire Detection Systems: A Survey, *Mohammad SamadiGharajeh Image and Vision Computing*, Vol. 91 (2019) 103803.

14. Tom Toulouse, Lucile Rossi, Turgay Celik and Moulay Akhloufi, Automatic Fire Pixel Detection using Image Processing: A Comparative Analysis of Rule-Based and Machine Learning-Based Methods, *Signal, Image and Video Processing* (2016) Vol. 10, pp. 647–654.

15. Multimedia Tools and Applications (2020) 79:9083–9099A Smart Approach for Fire Prediction under Uncertain Conditions using Machine Learning, Richa Sharma, Shalli Rani and Imran Memon, *Multimedia Tools and Applications*, 2020, Vol.79, pp. 28155–28168.

16. Lakshmisri Surya, Risk Analysis Model That Uses Machine Learning to Predict the Likelihood of a Fire Occurring at A Given Property, *International Journal of Creative Research Thoughts*, Vol.5, Issue 1, March 2017.pp. 959–962

17. T. Mikolov, M. Karafiát, L. Burget, J. Cernock ˇ ý,S. Khudanpur, Recurrent £. In *Proceedings of the Eleventh Annual Conference of the International Speech Communication Association*, Chiba, Japan, 26–30 September 2010, pp. 1097–1105.

18. M. Ahrens, *Trends and Patterns of US Fire Loss*, National Fire Protection Association: Quincy, MA, USA, 2017.

19. O. H. Kwon, S. M. Cho, S.M. Hwang, Design and Implementation of Fire Detection System. In *Proceedings of the Advanced Software Engineering and Its Applications*, Hainan Island, China, 13–15 December 2008, pp. 233–236.

# CHAPTER 6

# Structural Health Monitoring

Tanabalou Jayachitra[1] and Rashmi Priyadarshini[2]

*Sharda University, Greater Noida, Uttar Pradesh, India*

*E-mail ID: jayachitra.kishor@sharda.ac.in;*
*rashmi.priyadarshini@sharda.ac.in*

## 6.1 INTRODUCTION

In our day-to-day life we rely heavily on civil infrastructure, including buildings, bridges and so on. Some of these structures are older, therefore monitoring is important to avoid any mishaps. Structural health monitoring (SHM) is used to monitor structures continuously to detect any damage in the civil and mechanical structure using non-destructive methods. SHM is classified into two techniques: global and local. Severe damage falls under global techniques, which requires low frequency range. Local techniques are more sensitive to light damage, which work in KHz range. There are various methods using SHM, like the displacement method, vibration method and electromechanical impedance method. In these methods, the electromechanical impedance method is the most common since it uses a lead zirconate titanate (PZT) transducer. This transducer is used because of its small size, light weight and piezoelectric property. When the PZT is surface bonded with the structure, there is a direct relationship between the mechanical and electrical impedance of the transducer. The piezoelectric transducer acts as actuators and sensors. Damage in the structure is identified by measuring the electrical impedance method with proper excitation to the transducer in a suitable frequency range.

DOI: 10.1201/9781003217237-6

## 6.2 CLASSIFICATION OF STRUCTURAL HEALTH MONITORING BASED ON DIFFERENT STRUCTURES

Different structures are chosen for structural health monitoring. Structures like concrete beam, aluminum, stainless steel plates, portal frame structure and bolted joints were used.

Aluminium is chosen as a model for SHM.

The s structure chosen is an aluminium beam of $500 \times 30 \times 2$ mm. The damage in the structure is identified by attaching a steel nut of $14 \times 4$ mm and d 2g at various distances from the bonded transducer at the end of the beam. The frequency which was optimum in identifying damage is from 0 to 50 KHz. The voltage range in this frequency is 0.8 to 0.4 V. The impedance and Root Mean Square voltage values are obtained from the structure in undamaged condition and in damaged condition by keeping the nuts at a distance of 10cm, 20cm, 30cm and 40cm from the transducer. The RMS variation shows the sensitivity to damage [1–6]. Priya et al. conducted the experimental investigation on a homogenous plate made of aluminium with bonded PZT patches to investigate the external vibration at low frequency and boundary condition on electromagnetic impedance (EMI) signature [6]. Annamdas et al. presented an article which is a review of EMI technology with several materials like timber, concrete, aluminium, steel and so on, and issues like cracks, effects of loading, insertion of nails and moisture occurrence [7–14]. Baptista et al. investigated the impact of frequency excitation and voltage based on impedance. Tests were carried out on aluminum at different damage levels and the results demonstrate that the excitation frequency has a strong influence on the system performance and PZT dissipation of power [10].

The sensor node was tested on a frame with bolt joints. The impedance data were made with both the data acquisition and the suggested sensor node joined to the piezoelectric transducer patch with the frequency sweeping from 75 to 90 KHz. The maximum peak appears at 87 KHz under load condition and the bolt is fastened, then the maximum peak disappears [2].Experiments were conducted on a concrete beam subjected to differential curing by placing these bonded sensors on different thickness of steel plates. The frequency range is recorded from 100 KHz to 1000 KHz from day 2 to day28 [3].

Concrete structures were chosen for SHM [15–32]. The damages in concrete were indicated using displacement measurements from digital image correlation [3].Saravanan et al. presented a comparative study on the Electromagnetic Impedance (EMI) signatures from different smart aggregates embedded in concrete, during the starting stages of strength and cracking of the concrete surroundings [4].Concrete vibration sensors(CVS) are embedded inside a concrete beam [29,30–32]. The identification of damage was taken by various modes shapes (curvature and higher orders) [29].

Two steel structures are chosen with one healthy and the other damaged by placing two piezoelectric sensors to collect the data [13].Two piezoelectric sensors were mounted on the pin structure between the steel cable and the bridge deck [16]. A real-time flyover was designed using the finite element method [19].

YabinLiang et al. [33] placed the PZT sensors on the pin connected structures and damages were measured at different loading conditions. A typical SHM system is shown in Figure 6.1.

The PZT properties are shown in Table 6.1.

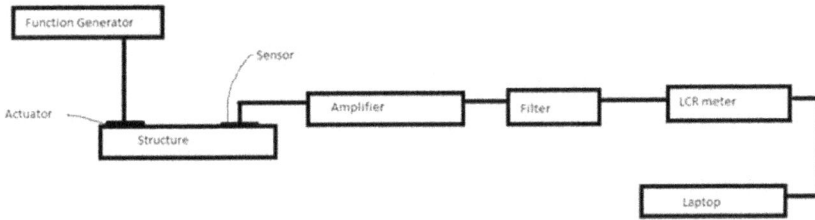

FIGURE 6.1   Block diagram of structural health monitoring.

TABLE 6.1   Properties of PZT (lead zirconate titanate patch)

| Parameters | Units | Values |
|---|---|---|
| Size (length × width) of the patch | M | $0.01 \times 0.01$ |
| Density | Kg/m³ | 7800 |
| Thickness of the patch | M | $2 \times 10^{-4}$ |
| Young's modulus | Pa | $46 \times 10^9$ |
| Dielectric loss | – | 0.02 |
| Mechanical loss | – | 0.0325 |
| Poisson ratio | – | 0.3 |
| Electric permittivity | $10^{-8}$ F/m | 1.75 |
| Piezoelectric strain coefficient | $10^{-10}$ mV$^{-1}$ | −2.10 |

## 6.3 VARIOUS TECHNOLOGIES USED IN STRUCTURAL HEALTH MONITORING

### 6.3.1 Impedance Measurement

The EMI technique is based on the piezoelectric effect, conversion of mechanical to electrical energy and vice versa. The RMS voltage of the piezoelectric sensor varies if the magnitude of EMI varies.

A novel method to detect the damage is based on the electromechanical impedance (EMI) principle for applications of SHM. The damage is quantified by the root mean square voltage from the piezoelectric transducer output signals. The impedance and RMS voltage values are measured from the structure in undamaged condition and in damaged condition [1–4, 11, 14, 20, 34, 35]. Impedance measurement is done by impedance analyzers. The impedance was costly so the impedance measurement was done by using direct digital synthesizer (DDS) and signal conditioning circuit. Now impedance measurement chip AD5933 is used for measuring impedance.

### 6.3.2 Admittance Measurement

The development of micro-cracks leads to damage in the form of cracks correlate with conductance obtained from electromechanical impedance measurements from the piezoelectric transducer, which is bonded to the concrete structure [4,5,21,36–38].

A novel method for damage identification based on electromechanical admittance of various PZT patches. The PZT patch functions simultaneously both as a sensor and actuator [7,17,29–32]. Admittance measurement measures real component conductance and imaginary component susceptance. The conductance shows active responses than the imaginary part susceptance.

## 6.4 SENSORS USED AND PLACEMENT OF SENSORS

The sensors used in SHM are divided into two categories: conventional sensors and smart sensors. Conventional sensors includes accelerometers, vibration strain gauges, electrical strain gauges and so on. Sensors have some specific properties which change the output according to the change in input. Smart materials include piezoelectric sensors, macrofiber composite transducers [35,36], shape memory alloys, optical fibres, electro-rheological fluids and magneto-strictive materials. Piezoelectric sensors are available in ceramic form (piezoelectric transducer, concrete vibration sensors and polymer (polyvinylidelene fluoride (PVDF)) [29].

## 6.5 METHODOLOGY

Cortez et al. [1] proposed a new method for detection of damage based on the electromechanical impedance (EMI) principle for SHM applications. The new SHM system is portable, autonomous and versatile. The damage is identified by the comparison of the root mean square voltage from the piezoelectric transducer output signals. The piezoelectric transducer, such as lead zirconate titanate patches, which are bonded to the structure are excited at various frequencies by DDS. The RMS voltage of the piezoelectric sensor varies if the magnitude of EMI varies. The sinusoidal wave generated by DDS is not proper so it must be applied to a conditioning circuit which has a low pass filter and buffer. The output of the transducer is given to the measurement circuit which converts AC to DC at each excitation frequency produced by DDS. The microcontroller used here is PIC16F877A, which controls the DDS device and its software. Ad5932 DDS and EEPROM 25LC1024, Butterworth low pass filter provide the excitation signal to the transducer with 2.3V amplitude. The current limiting resistor was chosen to be 1 KΩ. The data transfer between the microcontroller, DDS and external memory was done by serial peripheral interface. When data acquisition is done for the entire frequency, the RMS voltage obtained from the output of the transducer is compared with the reference signature. The root mean square deviation and correlation coefficient deviation metric is calculated from the root mean square voltage value. According to the intensity of damage detected, the threshold is set and in case of damage then the data is transferred to PC via RS232 interface. The structure chosen is an aluminium beam of 500 × 30 × 2 mm. Structural damage is identified by attaching a steel nut of 14 × 4 mm and d 2g at different distances from the transducer which is bonded at the end of the beam. The frequency which was optimum in identifying damage is from 0 to 50 KHz. The voltage range in this frequency is 0.8 to 0.4 V. The impedance and RMS voltage values are obtained from the structure in healthy condition and in damaged condition by keeping the nuts at a distance of 10cm, 20cm, 30cm and 40 cm from the transducer. These variations indicate that the RMS voltage is sensitive to damage.

Mascarenas et al. [2] developed a wireless sensor node with a low cost impedance measurement chip and radio frequency energy transmission to provide electrical energy to the sensor node for SHM. The impedance measurement analog devices are AD5933 and AD5934.They are embedded with analog to digital converter and digital to analog converter and fast fourier transform with a frequency range upto 100 KHz. The sampling rate of AD5933 is 1 Mega-samples per second and for AD5934 250 kilo-samples per second. Due to higher sampling rate AD5933 is used. The sinusoidal signal excitation is provided by a DD Sat the desired frequency and it is passed to the structure through the programmable gain stage. The current output is given to the current to voltage amplifier which has a feedback resistor. The signal will be in the linear range by choosing appropriate feedback resistor. Then it is sent to the analog to digital converter. The digital data is then sent through a hanning window and discrete fourier transform at the specified frequency. The resulting values are sent to the microcontroller. The microcontroller gives a command to the AD5933 to proceed to the next frequency point. The proposed sensor node was tested with a frame structure and a bolted joint. The impedance measurements were made with both the Data acquisition system and the proposed sensor node connected to the same piezoelectric transducer patch and swept frequency from 75 to 90 KHz. The dominant peak appears at 87 KHz under no preload and the bolt is fastened, then the dominant peak disappears.

Priya et al. [3] proposed a novel method on impedance based SHM using seven serially connected piezoelectric sensors. Experiments were conducted on a concrete beam subjected to differential curing by placing these sensors bonded on stainless steel plates of different thickness. The frequency is recorded in the range from 100 KHz to 1000 KHz from day 2 to day28.The excitations are provided to the patches and their admittance is measured using LCR meter, which is further connected to a laptop. The peak frequency is measured at different patches and their variations are compared with the series signatures taken base 2 as reference. The root mean square deviation is also calculated as damage index. The good curing side has higher conductance and an increase in root mean square deviation. The root mean square deviation of serial sensing has the same trend as the root mean square deviation of the individual patch. The serial sensing PZT method is more effective and time consuming method which has more applications in SHM applications.

Saravanan et al. [4] presented a comparison in evaluation of the Electromagnetic Impedance (EMI) signatures collected from five smart aggregates, which are embedded inside concrete, during the initial strength development and cracking of the surrounding concrete. The increase in the stiffness and the rightward shift of the resonant peaks and decrease of damping, with the consequent upward shift of amplitudes that happens during curing and strength gain, was observed to be reversed during the application of damaging loads.

Narayanan et al. [5] evaluated the damages in concrete using displacement measurements on the concrete cube obtained from digital image correlation. The origin of micro-cracks, leading to localization of damage in the form of cracks, is shown to correlate with conductance obtained from electromechanical impedance measurements from piezoelectric transducer bonded to the concrete structure.

Priya et al. [6] conducted the experimental investigation on a homogenous aluminium plate bonded with PZT patches to study the effects of low frequency vibration and boundary condition on electromagnetic impedance signature.

Dansheng Wang et al. [7] presented a new damage detection method based on electromechanical admittance of multiple PZT patches. The PZT patch simultaneously functions both as sensor and actuator. Baptista et al. [8] proposed a new impedance measurement method for an SHM system based on the EMI technique. The hardware consists of a low cost data acquisition (DAQ) device model and a laptop. The software used was LabVIEW, which is a graphical programming environment. The transfer of communication between the DAQ and the PC is performed through a USB port. The system offers high precision, speed, low cost and versatility, and can be used as a general impedance analyzer.

Annamdas et al. [9] presented an article which is an overview of EMI technology related to several materials (timber, concrete, aluminium, steel and so on) and issues (crack propagation, loading, nail insertion, moisture attack and so on) effecting these materials in a simple manner.

Baptista et al. [10] investigated the influence of the type of excitation signal and voltage level on impedance-based monitoring systems. Tests have been carried out on an aluminum specimen with different health conditions and the results show that the excitation signal has influence on system performance and power dissipation in PZT.

Nicolas et al. [11] presented the design and performance of a wireless based SHM system. Each sensor node is a mobile and autonomous SHM core based on a microcontroller, DDS and transceiver. The detection of damage is performed by the comparison of the changes in root mean square voltage received from the PZT, such as lead zirconate titanate patch, which is bonded to the structure.

Sreevallabhan et al. [12] presented a paper on SHM using wireless sensor networks. The ambient vibrations and strong motions are monitored by micro electromechanical systems accelerometer connected to the structure. The data are wirelessly transmitted and received at any remote locations. Two major components in wireless sensor networks are the transmitter board, comprising sensors, signal conditioning, microcontroller and wireless transmission, and the other receiver board, comprising receiver, data acquisition and microcontroller.ADXL335 accelerometer sensor is used due to its reduced power and 3 axis of orientation and ATMEGA 328P as processor. The piezoelectric sensor is chosen for its accuracy. The piezoelectric sensor is attached with the seismic mass having a string at the top. Whenever the vibration is produced through the string, the mass will experience the stress and piezoelectric crystal produces the proportional voltage. The transceiver transfers the information through radio signals. The vibrations are detected by the sensor and send the data to the microcontroller. The analysis of accelerometer was made with and without vibrations.

Abdelgawad et al. [13] proposed SHM utilizing the internet of things, which comprises Pro-Trinket microcontroller, transceiver module, Wi-Fi module and Raspberry pi. Two steel structures are chosen, with one healthy and the other damaged by placing two piezoelectric sensors to collect the data. One of the sensors gets the location of the damage and the other gets the size of it. Pro-Trinket is the Arduino ATmega328 used for implementing the proposed mathematical model to detect the damage. The nRF24L01+ module is used for its low power consumption and 2.4 GHz ISM band. It will transmit the data from the Pro-Trinket to the Raspberry Pi. The quadcore ARMCortex-A7 provides the data analysis from several nodes and sends to the cloud using Wi-Fi. Thingworx Platform is used for the internet of things. Damage identification is performed by Pro-Trinket and sends to the hub which is then uploaded in the cloud. The data in the cloud can be remotely checked from any mobile devices.

Annamdas [14] presented a paper on crack monitoring using the Electromechanical Impedance technique. Studies are held in two metallic specimens bonded with two piezoelectric transducers. Initially a small crack was produced and then it was allowed to propagate in steps. Root square deviation index was used to identify the crack propagating directions and numerical modal analysis for modal frequency shift.

Giurgiutiu et al. [15] presented high frequency SHM using the Electromechanical Impedance technique. Two sensors were used, one as sensing unit and the other as actuator unit. Impedance analyzer was used to measure the high frequency impedances and data calibration was performed using a microcomputer. The sensing and actuation was effective in the near field region, not in the far field region.

Yabin Liang et al. [16] reviewed on load monitoring of the pin-connected structure using time reversal technique and piezoceramic transducers. Two piezoceramic sensors were mounted on the pin connected structure between the steel cable and the bridge deck. One of the sensors is used as an actuator and the other sensor is used as the wave detector. The steel rod tension is adjusted by tightening the steel nuts for different loaded condition. In real time the tension of the rod was measured using optical fiber sensor. For practical applications, the proposed method was repeated for consistent results.

Swagato Das [18] reviewed the damage detection techniques by producing various vibrations used for health monitoring of structures. Modal analysis is the basic method if noise free areas are chosen, but in real life, situation distortions are present in the atmosphere and environment. The various methods in modal analysis are: frequency response method, strain energy method, and mode shape method. The frequency response analysis method and fuzzy clustering has to be integrated to reduce the noise and then the data are normalized. The strain energy method includes stiffness and mode shape matrix. If the finite element method is considered, this method is more effective to get the accurate results. Mode shape method includes the higher mode shapes, characteristics and higher order derivatives to identify the damage in the structure. In the local diagnostic method, piezo-electric sensors are used to detect damage by measuring the electric potential in time domain. Then the data are changed to frequency domain. The limitation in modal based analysis has been effectively countered. Based on the probability functions principle and modal parameters, many methods are defined to detect the damage. For monitoring of structural health, the Bayesian method used complex probability functions but probability density function is difficult to define accurately. Autoregressive models require greater numbers of accelerometers and complex calculations. Time series analysis model in combination with cloud computation can process large amounts of data effectively. In all these methods, accurate placements of sensors are very important. The conclusion was made that the time series method is a more effective method.

Kaur [19] proposed energy harvesting using thin piezo patches. The output voltage and power were generated from the bonded piezo in $d_{31}$ strain mode. A real-time flyover was designed using the finite element method. The patch was bonded to the plate and power generated was in the range of microwatts. Piezo patch bonded to I section beam. Excitation was provided by series portable dynamic shaker. The electrical signal was generated by a function generator and then amplified by a power amplifier. The dynamic shaker received the signal and produced mechanical force. The input excitation signal was in the range of 10 to 100 Hz at 5V. The accelerometer was attached in the center of the beam to measure its amplitude. The mechanical losses, dielectric losses and shear lag effect were also measured.

Perara et al. [20] identified crack debonding in reinforced concrete beam with fiber reinforced plastic. A multi-objective model is obtained by the various impedance measurements for several sensors and tested at various levels of damage. Instead of the finite element method, the spectral element method was used. Generally, the objective function is written from the differences between the numerical and experimental measurements. Impedance was measured from the sensors and root mean square deviation was calculated to make single objective function. Multiple objective function methodology has more advantages than single objective function, like modeling errors and differences in the data measured. Particle swarm optimization method was preferred to form multiple objective functions.

Kaur et al. [21] evaluated experimentally the low cost and small size impedance chip for SHM. The experiments were performed in RC beam and mild steel. Many pairs of piezoelectric ceramic lead zirconate titanate patches were embedded in a real-life RC beam. First-level damage can be identified using an impedance chip, which is more precise and has an inbuilt function generator, temperature sensor. Power for the chip can be provided by the computer's USB port. The system clock frequency is 16 MHz. The accuracy is 0.5% in the frequency range from 1 to 100 KHz. Calibration of the impedance chip is required each time during reset of hardware or software. It can be removed by adding 200KΩ resistance and running the evaluation software as per the instructions in the manual. Aluminium strip was chosen for test and a hole was created by drilling and the impedance was measured by using both impedance chip and LCR meter. Then the experiment was tried in RC beam by placing CVS in the steel plate. The conductance value in both AD5933 and LCR meter has the same trend but different magnitude due to the proper sinusoidal signal obtained in LCR meter. Impedance chip and self sensing macrofiber composite patch was used to detect the corrosion in an aluminium structure [35]

Negi [22] investigated the effectiveness of embedded PZT patches at different orientations ($0^0, 45^0, 90^0$) during monitoring the hydration of RC beam. The dynamic shaker was used to actuate the beam. In all the PZT, the peak voltages at different frequencies were measured. The least response was found in 45 degree orientation compared to the other two. The PZT patch horizontally placed detects the damage better than the other orientations.

Kaur et al. [23] presented a comprehensive study on four types of configurations, surface bonded single piezo transducers, embedded single piezo transducers and metal wire single piezo transducers. Kaur et al. and Bhalla and Kaur [29,30] used Concrete vibration sensor were embedded in the RC structures and measured the damage with respect to stiffness of the structure. The CVS and surface bonded PZT's output voltage and power were compared and found the cracks were identified effectively in CVS [34]. Negi et al. [31] used CVS embedded in different orientations and found cracks are identified in the horizontal position of sensor more accurate than the other inclinations. Kaur et al. [32] made the comparison between different ways of placing the sensors in the structure: surface bonded, embedded and metal wire configuration. The best crack detection was by metal wire configuration.

Yabin Liang et al. [33] placed the PZT sensors on the pin connected structures and damages were measured at different loading conditions using a 3D finite element model. Excitation signal was given at high frequency (>30 KHz) and damages were detected at various load conditions. The dominant frequency peak increases as the tension in the steel rod increases.

Usually the frequency range of the transducer is taken by a trial and error method to identify damage. Baptista et al. [36] proposed the range of frequencies in which the PZT are sensitive. Damages are developed at a distance of 100 mm from PZT at the frequency range from 0–125KHz.If the mechanical impedance of the structure is more than the impedance of the transducer, then the sensitivity of the transducer is affected. The frequency range of the PZT depends on type and damage of the structure.

PZT patches limitation in SHM [37]:

1. PZT patch material is brittle, therefore PZT cannot be placed in the bending surfaces of the structure.

2. PZT electrical properties fluctuate more with temperature.

3. Proper soldering wires with PZT patch and using proper adhesives increases the sensitivity of the transducer.

4. PZT erodes naturally after long period of time.

5. When applied to high excitation signal, the PZT patches depolarize.

6. Dipole alignment changes when there is increase in mechanical stress.

7. Depolarization takes place if PZT is heated above the curie point.

Due to the above limitations, there were some drawbacks in using PZT patches.

1. The sensing region becomes very small (metal 2m and non metal<0.2m).

2. The overall structure strength cannot be identified by small crack damage.

## REFERENCES

1. Cortez, Nicolás E., Filho,Vieira Jozue and Baptista, Guimarães Fabricio (2012), "A new microcontrolled structural health monitoring system based on the electromechanical impedance principle", *Structural Health Monitoring* 12(1):14–22.
2. Mascarenas, David, L., Todd, Michael D., Park, Gyuhae and Farrar, Charles, R. (2007), "Development of an impedance based wireless sensor node for structural health monitoring", *Smart Materials and Structures* 16: 2137–2145.
3. Priya Bharathi, C., Gopalakrishnan, N. and Rao Mohan Rama, A. (2015), "Impedance based structural health monitoring using serially connected piezoelectric sensors", *Journal of Institute of Smart Structures and Systems (ISSS), JISSS* 4(1):38–45.
4. Saravanan Jothi T., Balamonica K., Priya Bharathi C., Reddy Likhith A. and Gopalakrishnan, N. (2015), "Comparative performance of various smart aggregates during strength gain and damage states of concrete", *Smart Materials and Structures* 24 085016.

5. Narayanan, Arun and Subramaniam, Kolluru V. L. (2016), "Sensing of damage and substrate stress in concrete using electro-mechanical impedance measurements of bonded PZT patches", *Smart Materials and Structures* 25 095011.

6. Saravanan Jothi, T., Balamonica, K., Priya Bharathi, C., Reddy Likhith, A. and Gopalakrishnan, N. (2014), "Low frequency and boundary condition effects on impedance based damage identification", *Case Studies in Nondestructive Testing and Evaluation* 2: 9–13

7. Wang Dansheng, Song Hongyuan and Zhu Hongping (2013), "Numerical and experimental studies on damage detection of a concrete beam based on PZT admittances and correlation coefficient", *Journal of Construction and Building Materials* 49:564–574.

8. Baptista, Guimarães Fabricio and Filho,VieiraJozue (2009), "A new impedance measurement system for PZT-based structural health monitoring", *IEEE Transaction on Instrumentation and Measurement* 58(10):3602–3608.

9. Annamdas, Venu G. M. and Radhika, Madhav A. (2013), "Electromechanical impedance of piezoelectric transducers for monitoring metallic and non metallic structures: A review of wired, wireless and energy-harvesting methods", *Journal of Intelligent Material Systems and Structures* 24(9):1021–1042.

10. Baptista, Guimarães Fabricio, Filho, Vieira Jozue and Inman, Daniel J. (2010), "Influence of excitation signal on impedance-based structural health monitoring", *Journal of Intelligent Material Systems and Structures*, 21: 1409–1416.

11. Cortez, Nicolás E., Filho, Vieira Jozue and Baptista, Guimarães Fabricio (2015), "Design and implementation of wireless sensor networks for impedance-based structural health monitoring using ZigBee and global system for mobile communications", *Journal of Intelligent Material Systems* 26(10): 1207–1218.

12. Sreevallabhan B., Nikhil Chand, B. and Sudha, Ramasamy (2017), "Structural health monitoring using wireless sensor networks", *IOP Conf. series: Materials Science and Engineering* 263 052015.

13. Abdelgawad, Ahmed and Yelamarathi, Kumar (2016), "Structural health monitoring: Internet of things application", *2016 IEEE 59th International Midwest Symposium on Circuits and Systems (MWSCAS)*, 16–19 October 2016, Abu Dhabi, UAE.

14. Annamdas,Venu Gopal (2012), "Facts of impedance technique in crack propagation studies for an engineering structure", *International Journal of Aerospace Sciences*:8–15.

15. Giurgiutiu,V. and Rogers, C.A. (1997), Electro-mechanical (E/M) impedance method for structural health monitoring and non-destructive

evaluation, in: *International Workshop on Structural Health Monitoring*, Stanford University, CA, pp. 434–444.

16. Liang, Yabin, Li, Dongsheng, Kong, Qingzhao and Song, Gangbing (2016), "Load monitoring of the pin-connected structure using time reversal technique and piezoceramic transducers: A feasibility study, *IEEE Sensors Journal*, 16(22).

17. Dansheng Wang, Hongyuan Song and Hongping Zhu (2014), "Embedded 3D electromechanical impedance model for strength monitoring of concrete using a PZT transducer", *Smart Materials and Structures* 23:115019 (14pp).

18. Das, Swagato, Saha, P. and Patro, S. K. (2016), "Vibration-based damage detection techniques used for health monitoring of structures: A review article", *Journal of Civil Structural Health Monitoring* 6:477–507.

19. Kaur, Naveet and Bhalla, Suresh (2014), "Feasibility of energy harvesting from thin piezo patches via axial strain (d 31) actuation mode, *Journal of Civil Structural Health Monitoring* 4(1):1–15.

20. Perara Ricardo, Sun Rui, Servillano, Enrique and Ruiz, Antonio (2017), "A multi-objective electromechanical impedance technique to identify debonding in RCbeams flexural strengthened with FRP", *X International Conference on Structural dynamics* 2232–2237.

21. Kaur, N., Bhalla, S., Shanker, R. and Panigrah,i R. (2015), "Experimental evaluation of miniature impedance chip for structural health monitoring of prototype steel/RC structures", *Experimental Techniques, Society for Experimental Mechanics.* 40:981–992.

22. Negi, P., Chakraborty, T., Kaur, Naveet and Bhalla, Suresh (2018), "Investigation of effectiveness of embedded PZT patches at varying orientations for monitoring concrete hydration using EMI technique", *Construction and Building Materials* 169:489–98.

23. Kaur, Naveet, Li, Lingfang, Bhalla, Suresh, Xia, Yong, Ni, Pinghe and Adhikari, Sailesh (2017), "Integration and evaluation of multiple piezo configurations for optimal health monitoring of reinforced concrete structures", *Journal of Intelligent Material Systems and Structures*, 28(19):2717–2736.

24. Yao, J.T.P (1985), "Safety and reliability of existing structures", *Pitman Publishing Programme*, London.

25. Oreta, A.W.C. and Tanabe,T. (1994), "Element identification of member properties of framed structures", *Journal of Structural Engineering, ASCE* 120(7):1961–1976.

26. Loh, C.H. and Tou I.C. (1995), "A system identification approach to the detection of changes in both linear and nonlinear structural parameters", *Earthquake Engineering & Structural Dynamics* 24(1):85–97.

27. Adams, R.D., Cawley, P., Pye, C.J. and Stone, B.J. (1978), "A vibration technique for non destructively assessing the integrity of structures", *Journal of Mechanical Engineering Science* 20:93–100.

28. Zimmerman, D.C. and Kaouk, M. (1994), "Structural damage detection using a minimum rank update theory", *Journal of Vibration and Acoustics* 116:222–231.

29. Kaur, Naveet, Bhalla, Suresh and Maddu, Subhash C.G. (2019). "Damage and retrofitting monitoring in reinforced concrete structures along with long-term strength and fatigue monitoring using embedded lead zirconate titanate patches", *Journal of Intelligent Material Systems and Structures*, 30(1):100–115.

30. Bhalla, S. and Kaur, N. (2018), "Prognosis of low-strain fatigue induced damage in reinforced concrete structures using embedded piezo-transducers", *International Journal of Fatigue* 113: 98–112.

31. Negi, Prateek, Kaur, Naveet, Bhalla, Suresh and Chakraborty, Tanusree (2015), "Experimental strain sensitivity investigations on embedded PZT patches in varying orientations", In V. Matsagar (ed.), *Advances in Structural Engineering*, DOI 10.1007/978-81-322-2187-6_203, Springer India.

32. Kaur, Naveet, Li, Lingfang, Bhalla, Suresh and Xia, Yong (2017), "A low-cost version of electro-mechanical impedance technique for damage detection in reinforced concrete structures using multiple piezo configurations", *Advances in Structural Engineering* 20(8):1247–1254.

33. Liang, Yabin, Li, Dongsheng, Parvasi, Seyed Mohammad and Song, Gangbing (2016), "Load monitoring of pin-connected structures using piezoelectric impedance measurement", *Smart Materials and Structures* 25, 105011 (14pp).

34. Kaur, Naveet and Bhalla, Suresh (2014), "Combined energy harvesting and structural health monitoring potential of embedded piezo-concrete vibration sensors", *Journal of Energy Engineering*, 141(4):1–18.

35. Park, Seunghee, Grisso, Benjamin L., Inman, Daniel J. and Yun, Chung-Bang (2007), "MFC-based structural health monitoring using a miniaturized impedance measuring chip for corrosion detection", *Research in Nondestructive Evaluation* 18: 139–150.

36. Baptista, F. G., Filho, J. V. and Inman, D. J. (2010), "Influence of excitation signal on impedance-based structural health monitoring", *Journal of Intelligent Material Systems and Structures* 21(14), 1409–1416.

37. http://web.iitd.ac.in/~sbhalla/thesispdf/visalakshi.pdf.

38. Shanker, R., Bhalla, S. and Gupta, A. (2010), "Integration of electro-mechanical impedance and global dynamic technique for improved structural health monitoring", *Journal of Intelligent Material Systems and Structures* 21(2):285–295.

# CHAPTER 7

# Application of Machine Learning in Agriculture with Some Examples

Amrita Rai[1] and OM Prakash[2]

[1]*SMIEEE & Associate Professor, GL Bajaj Institute of Technology and Management Greater Noida, India*

[2]*Associate Professor, Sri Venkateswara College of Engineering and Technology, Chittoor, India*

E-mail: *amritaskrai@gmail.com; om4096@gmail.com*

## 7.1 INTRODUCTION: BACKGROUND AND MOTIVATION

For survival, more than 60%of the world's population depends on agriculture, according to the Food and Agricultural Organization of the United Nations (2015). According to The Indian Economic Survey 2017–18, more than 50%of the Indian population depends on agriculture. That means more than 60% of the people of India depend on agriculture and agrobased industries for their livelihood.

The major problem with agriculture is that it is highly sensitive to climate as well as soil. The recent technique of deep learning constitutes a modern method for image processing and data analysis with promising results and enormous potential. Having been successfully applied in various fields, deep learning has recently entered the area of agriculture [1]. Sixteen significant areas have been identified where we generally use deep learning, such as: land cover classification, crop type classification, fruit counting, weed detection and plant recognition. The overall benefits

DOI: 10.1201/9781003217237-7

**139**

of deep learning are encouraging for its further use towards smarter, more sustainable farming and more secure food production [2].

Agriculture is the backbone of economic growth in developing countries. The quality of production depends on the healthy growth of plants. Disease detection in early stages can improve the production percentage. Recently, various image processing techniques have been developed to detect leaf diseases. Plant diseases occurs when an organism infects a plant, leaf, or fruits and disrupts normal growth habits. The proposed technique helps in identification and detection of plant disease, which can be used as a defense mechanism to detect against the disease. Presently, the loss of food is mainly due to infected crops, which reflexively reduces the production rate. Techniques to enhance precision agriculture which might be primarily based on the images of plants or datasets are used to locate exceptional diseases in plants. This is accomplished to lessen the financial losses in agricultural department. The use of the correct facts is an essential difficulty of every system. Several diseases that affect tomato plants and their characteristics are the main part of this study [3]. This chapter discusses the technique based on digital image processing, which has been utilized for the detection and classification of leaf disease present on different agriculture plants. In this chapter, we analyzed and presented the result of various sorts of tomato plant diseases. Next to it, different techniques used earlier including computational intelligence is provided. Further, use of deep neural network as an emerging intelligent technique is provided for a different mode of analysis by researchers. Finally, an example of classification is depicted with future prospects [4].

In India, agricultural crop season starts in July and ends in June. This season is divided into two parts. According to monsoon the first is Kharif and second is Rabi. Kharif and Rabi are word is Arabic words where the meaning of these words are autumn and spring, respectively. Kharif crops start in July and Rabi crops in October to March, which is the winter season. Rabi crops are wheat, chickpea, oat, barley, peas, sunflower, and so on, and Kharif season crops are bajra, jowar, maize, soybean, and so on. Raining seasons depend on the temperature and potential to evapotranspiration. Across the world, some areas are better for the growth of crops. According to the equator, the world has different seasons depending on the distance of countries. In the season of the wheat crop, the growing season causes market prices to fluctuate. In the United States and China are two types of wheat; the first is winter wheat and the second is spring wheat. China is one of the biggest producers of wheat compared to other countries. After China, the biggest producers are India and the United States.

## 7.2 CLASSIFICATION FOR AGRICULTURE

One of the world's most prevalent activities is "agriculture" – the cultivation of certain plants and the raising of domesticated animals use the process of producing foods, fiber and some other products. But this is not the same in all countries. "Farming" is the word is used for the agricultural process. In agricultural working, processes encountered many changes in the twentieth century, especially in the machine learning techniques used in agricultural forecasting. Some criteria and classification are adopted for agriculture. Classification of agriculture is broadly categorized in two parts: commercial and subsistence farming. Further, these criteria are sub-classified into many different parts as shown in Figure 7.1.Subsistence agriculture happens when a plot of land creates just sufficient food to take care of the family working it or the neighborhood local gathering (clan and so forth), settle charges and here and there leave some sort of surplus to bargain with or to sell in better years. Shifting cultivation is a good example of subsistence agriculture, whereas commercial agriculture happens on

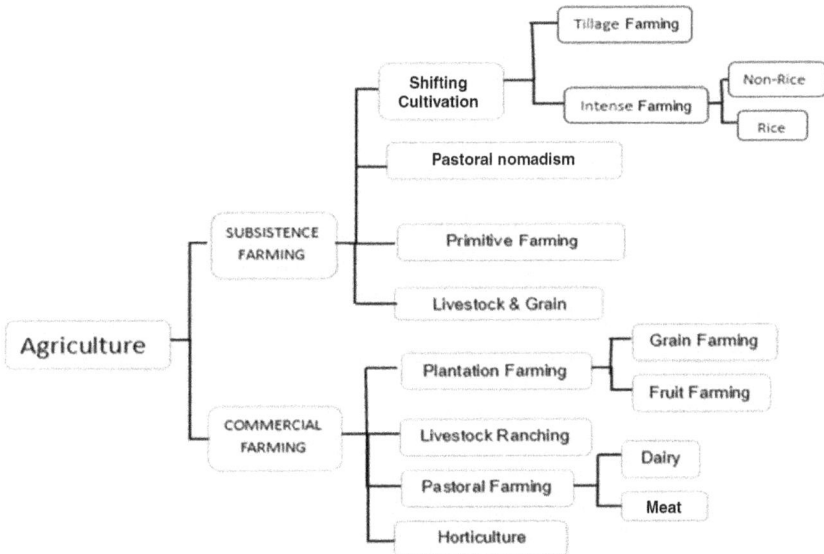

FIGURE 7.1   Classification and type of agriculture world wide.

an enormous, benefit-making scale [5]. It might very well be done by singular ranchers or by organizations, with the two gatherings attempting to increase the profit from information sources and looking for the highest yields per unit of land. Agriculture with the market grading such as cattle ranching, commercial grain farming, and the intensive cultivation of fruits, flowers and vegetables are examples of commercial farming.

Agriculture can also be classified on the basis of study of different fields related to farming and nature. Figure 7.2 gives brief details of different branches used in agriculture: agronomy, horticulture, plant breeding/genetics, soil science, agro-meteorology, agricultural biotechnology, agricultural engineering, agricultural extension and so on.

Different branches require different types of study and technology, for instance, horticultural plantation needs systematic improvement by application of genetic principals, and fishery science deals with the discipline of aquaculture where it studies aquatic resources, medical treatment and diagnosis of diseases in fish and aquatic animals. Similarly, every branch has their own significance and requirement in agriculture worldwide. In the field of agriculture the yield of the crops important. For a better yield of crops, farmers are using a yield prediction. This is the most important and popular topic in precision agriculture. It is represented by yield mapping and estimation of yield, comparing the supply and demand of crop and crop management. According to past data, we are moving far away from simple predictions by using technologies to makes the most of the yield for farmers and populations. These technologies include artificial intelligence

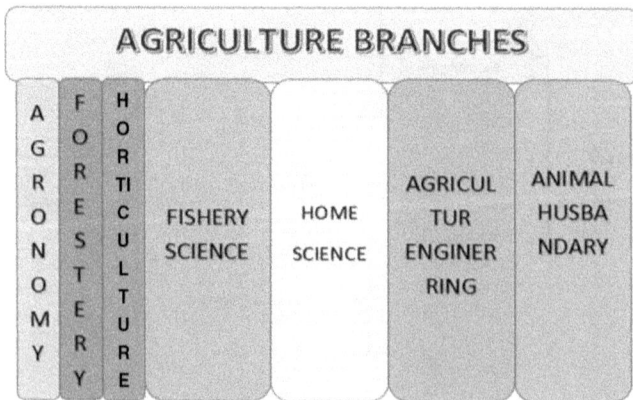

FIGURE 7.2   General branch of farming and agriculture.

(AI), machine learning (ML) and Neural Network. Fundamental for seasons of farming is the space of one year. The farming season has fundamentals for better yields of crop like-wise, temperature and moisture as conditions are suitable for better yield of the crop. That helps farmers to arrange the yield production. Under farming season, climate change especially helps farmers for farming plant and growing crops.

## 7.3 TECHNOLOGY IN AGRICULTURE

Technology has a vast role in the agricultural industry. The innovation of technology is able to modernize this industry with complex solutions leading to better yields. Many sensors, machinery and tools have helped farmers to play a vital part in developing the economy. Figure 7.3 show that the different technologies used in the development of agriculture and cultivation.

Technological innovations and implementation in the field of agriculture have many impacts, as depicted in Figure 7.4.

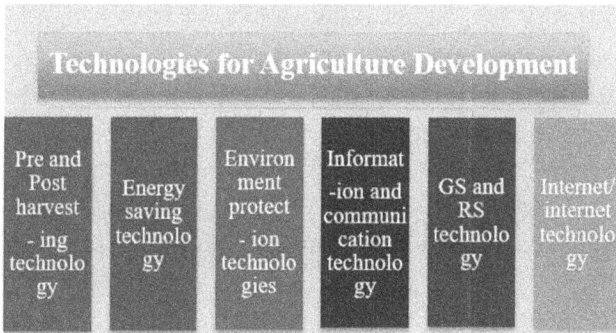

FIGURE 7.3    Various technologies for agriculture development.

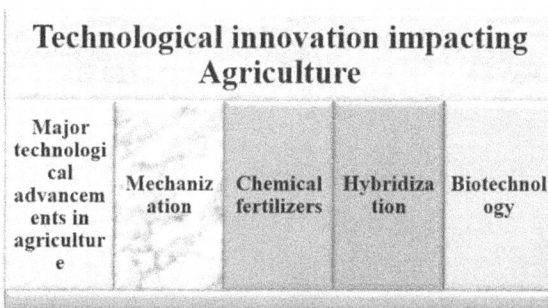

FIGURE 7.4    Impact of technological innovation in different field of agriculture.

The main areas of innovation are mechanization farming, chemical fertilization, biotechnology and fishery sciences. The field of technological innovation in agriculture was revolutionized after the introduction of the Internet of Things (IoT) [6], machine learning, deep learning, data science and artificial engineering. In this chapter, the basic concept of machine learning and its utilization in agriculture field is demonstrated with some examples [7]. Data analysis, machine learning and artificial intelligence have one common and important property, that is, "data". If we don't have data, then we can't train a model. Big enterprises spend money in large quantities to collect data. This can be in any form, such as value, text, picture, and so on, to be described and analyzed. We can divide the data for analysis into three parts for machine learning, which are: training data, validation data and testing data. After learning the need and type of data processing is most important [8].In this process, the concept of machine learning arises.

## 7.4 MACHINE LEARNING STRUCTURE FOR AGRICULTURE

Machine learning is a multidisciplinary field, which combines computer science and statistics, and is mostly used for analysis, classification and performing the tasks that humans generally do. For that we need to train the computer to solve real-life problems with the utmost accuracy. Figure 7.5 represents a working model of machine learning. Machine learning allows the system the capability to automatically explore, enhance and improve from the different experiences without being programmed [9].

The basic components of machine learning are statistics, calculus, linear algebra and probability. Machine learning centers on the development of

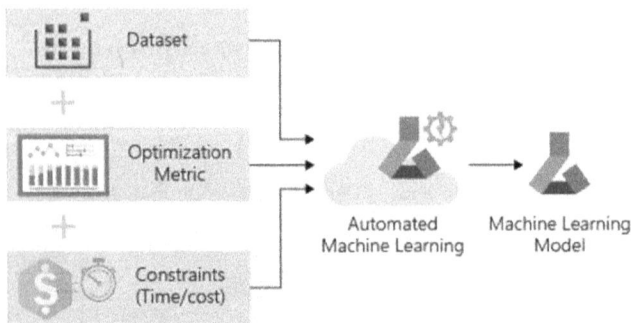

FIGURE 7.5   Working model of machine learning.

intelligent computer programs that can way the data and utilize it to learn from them. Machine learning is powered by the diamond of statistics, calculus, linear algebra and probability statistics at the core of everything [10]. Calculus tells us how to learn and optimize our mode; linear algebra makes running these algorithms feasible on the massive data sets; and probability helps predict the likelihood of an event occurring. Today's world is surrounded by many technologies for agriculture. So it is important to stay up-to-date with new emerging technologies using machine learning for agriculture [11].

## 7.5 DIFFERENT ALGORITHMS USED IN MACHINE LEARNING

The machine learning model obtains experience overtime, then improves its performance. Some statistical and mathematical models are used to determine the performance of the machine learning model and machine learning algorithms. When the learning process is completed then the model may be used to classify and make the assumption, and to test data. It can be achieved after the training process is completed [12]. Basic algorithms of machine learning are: supervised machine learning algorithm; unsupervised machine learning algorithm; and reinforcement machine learning algorithm [13].Supervised learning is a method used to enable machines to classify/predict object problems or situations based on laved data fed to the machine. In unsupervised learning, a machine learning model finds the hint for the unlabeled data.

## 7.6 APPLICATIONS OF MACHINE LEARNING IN AGRICULTURE

Uses of machine learning in different areas of agriculture are shown in Figure 7.6.

AI is uses in various sectors, from home to offices, all over the world for different purposes and presently also in agriculture. In the agricultural field, the machine learning technique used increases productivity and the quality of the crop [14]. Some of the applications used in agriculture are outlines below.

**Soil Management:** This plays a key role in yield efficiency, ecological stability and human health, both directly and indirectly. Soil is a diverse natural source, having complex processes and fuzzy mechanisms in

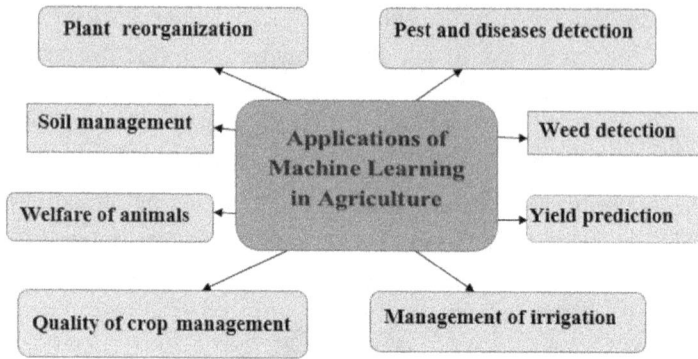

FIGURE 7.6    Major areas of agriculture where machine learning applies.

which the temperature of soil also plays an important function in the precise investigation of climatic variations of an area and its ecological behavior. Machine learning algorithms play a significant role in analyzing the soil temperature and dampness so as to understand the dynamics of ecosystems and its impact on agriculture. Goap et al. [15] have introduced the new technique back propagation networks, which gives better results for finding the beneficial properties of soil, rather than using the traditional method called multivariate regression model. The working rule of back propagation networks is to train a particular crop which has certain properties.

**Plant Reorganization**: In comparison to the conventional approach for classification of plants by comparing shape and color of leaves, machine learning can give exact and faster results to analyze the leaf vein morphology, carrying additional information leaf of characteristics. The foremost objective is the automatic recognition and categorization of different plant varieties so as to avoid the need for the human expertise, and also to minimize the categorization time. Bu. F & Wang. X [16] used deep convolutional network for the problem of plant identification from leaf vein patterns. They considered 3 legume species – white bean, red bean and soya bean leaf vein patterns; where vein morphology carries out the information of the leaf, it is one of the major tools for plant identification in comparing color and shape .In paper [17] the authors have modelled a methodology to differentiate the species of plant using a low resolution 3D lidar sensor. The authors have modelled a feature set having common statistical features being independent of plant size. The classifiers have been trained and compared in this model with the feature set that shows high efficiency in identification.

**Pest and Disease Detection:** Pest and disease control is one of the main problems in today's agriculture. One of the methods to control disease and pests is to uniformly spray pesticides over the crops, which results in high efficiency but it is not economical; it also leads to side effects of ground water and impact on wildlife and ecosystems. In paper [18] authors have developed a machine which helps to identify parasites in the greenhouse environment through image processing. The support vector machine (SVM) method can be used for classification and targeting parasite detection. The image processing methodology and SVM method, having the appropriate option of province and color index, proved to be successful for detection of objectives. With high efficiency, it has developed an effortless and economical optical gadget for remote ailment exposure, based on awning reflectance in numerous wavebands [19, 20]. They investigated the difference between healthy and ailing plants with early stages of yellow rust; field images were occupied by placing a spectrograph at spray resonant point and using intensity normalization to decrease the spectral high variability caused by canopy architecture and at different illumination levels. Aquadratic discriminating model based on the reflectance of these wavebands will classify healthy and disease spectra with a high success rate.

**Weed Detection:** For good yield production, prevention of weeds is one of the major issues. Weed detection and prevention is difficult to discriminate from crops, so machine learning uses sensors, which leads to precise detection and prevention of weeds with less expenditure and also does not harm the environment. The authors in paper [21] used remote sensing for discrimination of species and for operational weed mapping, exposure and mapping of silybrum-marianum weed patches by means of ordered self-organizing map reported using a multivisionary camera, which gives high resolution images.

**Yield Prediction:** There are many factors by which farmers can get optimum results in agriculture. One of these factors is to predict the yield of crop. This includes the fertility of the soil, irrigation process, climate conditions and controlling of pests. If the farmers do not follow these four factors correctly during farming there is a huge risk of damage to crops. Let us see some of the machine learning models which are used in the agriculture sector. There is a machine learning application which helps to count the number of coffee seeds in a branch and also segregates the coffee fruits in three categories such as harvest, non-harvest and seeds with disregarded maturation stage. Also we can estimate the weight of seeds and maturation percentage of coffee seeds. In paper [22] it is shown how to count coffee fruit automatically from coffee tree using machine vision system. When the development of a crop is at an initial stage and the harvesting is not done, using MVS technique they prove that estimation of seed count will not be too high or too low and by this they have shown that it obtains a higher correlation value of 0.90.In paper [23] an MVS was developed which automatically shakes and catches the cherry during harvesting stage; it also detects the occluded branches and cherries which are not clearly visible, whereas during cherry harvesting more laborers are required, which takes around 50% of the annual production cost. To reduce this cost, mechanized harvesting technologies have been used, such as limb-actuators that vibrates the cherry fruit so that it can be released from the branches. This tool generates a new era in the horticulture sector, which has higher efficiency and is economical for farmers, so that they can use this technique for their agricultural work to increase their productivity

**Welfare of Animals:** The field of animal welfare takes care of the health and well-being of animals so as to create a balance in the ecosystem, with the key application of machine learning in monitoring animal behavior for the early exposure of infection. Khanna and Kaur [24] showed that they followed a 2-stage machine learning frame-work which is an effective method for classification of cattle behavior. Cattle sensor technology and assemble classifiers are used in a current approach to categorize and examine the behavioral changes for improving additional feed to each cow. Mohapatra and Mohanty [25] proposed a technique based on data collected by optical fibrebragg grating sensors, which are projected by machine learning technique (pattern classification). In this study they have considered chewing process and food intake of dietary supplement. They showed that pattern classification differentiates the five different patterns involved in the chewing process.

**Quality of Crop Management:** To increase the value of crops and reduce the wastage one has to classify quality of crop with minimum error. The penultimate sub-category for the crop is developed for the identification of characteristics associated with the crop class. Zhang et al. [26] developed a model for detection and classification of botanical and non-botanical foreign material rooted within cotton lint at the time of the harvesting process.

**Management of Irrigation:** Irrigation is an important part of agriculture. It should be neither too high nor too low but it should be balanced between the two. To maintain these conditions, certain factors need to be considered, which are: soil type, land topography, weather, type of crop, water quality and so on. Kuma et al. [27] discuss neuro drip, an Excel based artificial neural network designed to provide rapid illustration of soil wetting patterns from surface drip irrigation emitters.

## 7.7 COMPANIES ASSOCIATED WITH THE AGRICULTURE SECTOR

Agriculture is regarded as the face of the revolution, thanks to increasing technical improvement. Modern farming methods are used by farmers all over the world to improve agricultural harvests, while also protecting crops from droughts and encouraging innovation. Automation, robotics, and sophisticated analytics are reshaping industries' futures, making them more sophisticated and intelligent. Machine learning applications shine like a halo, allowing for fewer equipment failures, improved on-time deliveries, quality improvements, shorter training cycles, and advanced automation of design and production processes [28].

By 2050, the agriculture sector will be faced with enormous challenges in feeding 9.6 billion people. Agriculture will need to expand its capacities due to a restricted amount of land and water [29]. This is the point at which technology takes center stage, establishing a presence and better utilizing resources. Here are a few companies that have built their businesses on machine learning, leveraging the technology to gain a competitive advantage. Figure 7.7 shows one of the example of drone use in agriculture provided by company NatureSweet.

**NatureSweet:** NatureSweet LTD is a produce grower, packager and vendor situated in San Antonio. Glorys, Cherubs, SunBursts, Jubilees and Eclipses are among the tomato varieties grown by the company. The company is vertically integrated. NatureSweet grows tomatoes in greenhouses in Willcox, Arizona and in 6 Factories in Jalisco, Nayarit and Colima, three states in

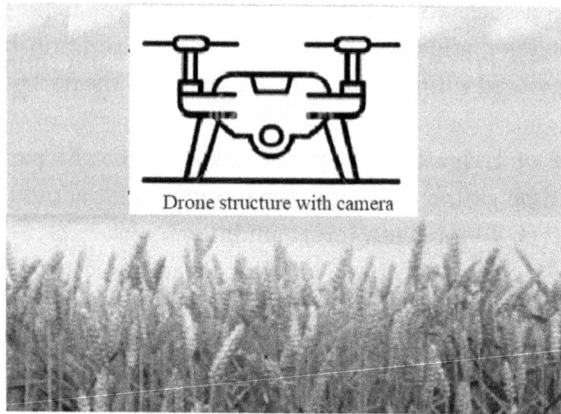

FIGURE 7.7    Drone system for agriculture.

Central Mexico. The organization monitors all types of emergent plant problems with the use of machine learning and algorithms in place, as well as cameras installed. "NatureSweet claims that machine learning analysis has increased harvest by 2 to 4%, with the founders hoping to eventually reach 20%" (https://revolveai.com/ai-applications-in-agriculture/).

**John Deere:** Blue River Technology was acquired by John Deere, a thriving 180-year-old corporation recognized for its characteristic green tractors. The food and agriculture chain, according to John Deere, is one of the most promising industries where machine learning may bring about dramatic changes.

He believes that machine learning will minimize the need for herbicides by nearly 95%, because computer vision will enable the agricultural revolution to make decisions about every single crop in the field. Machine learning will improve every aspect of agriculture, including crop harvesting, weather forecasting, soil tilling, specialized area selection, fertilizer usage and rainfall patterns.

**Bowery:** Bowery is a modern farming operation dedicated to producing the finest vegetables possible. Bowery chose to grow indoors instead of outdoors because huge farms are vulnerable to storms and rain. The startup's method involves using machine learning, LED lighting, data analytics and other techniques to grow leafy greens indoors with no pesticides and very little water.

**Plenty:** Plenty, one of the most influential indoor farming firms building the global food chain, announced a $200 million funding round spearheaded by the SoftBank Vision Fund, making it one of the largest agri-tech investments in history.

**Camposeven:** Camposeven teamed with Ec2ec to assist them in applying machine learning to improve the seed selecting process. Growing, harvesting and processing in distribution facilities to clients are all subject to stringent controls to guarantee that each product is of the greatest quality possible.

**Blue River Technology:** Weed control is one of their specialties. Automation and robotics are being used by companies like Blue River Technology to find more efficient ways to protect crops from weeds. The business created the See & Spray robot, which uses computer vision to accurately monitor and spray weed among cotton plants. Herbicide resistance can be avoided by spraying with precision. According to reports, the company says that their precise technology reduces herbicide expenditure by 90% and eliminates roughly 80% of the volume of chemicals sprayed on crops.

**Harvest CROO Robotics:** Crop harvesting is one of their specialties. CROO Robotics created a robot to aid strawberry farmers in picking and packing their products. Millions of dollars in revenue have been lost in key farming regions of the United States, notably California and Arizona, due to a lack of laborers. According to the business, its robot can harvest 8 acres in a single day and replace 30 human laborers.

**PEAT:** Specializes in the use of machine vision to detect pests and soil defects. PEAT is a Berlin-based technology company that uses deep learning to detect probable soil flaws and nutrient deficits. Plantix is the name of a deep learning-powered image recognition application. It uses photos to detect potential flaws and provides users with soil restoration strategies, advice and other useful information. Plantix, according to PEAT, can instantly recognize patterns with a 95%accuracy rate.

**SkySquirrel Technologies Inc:** Drones and crop analysis using computer vision are two of their specializations. SkySquirrel Technologies Inc. is bringing drone technology to vineyards. Its mission is to help farmers boost agricultural yield while cutting costs. Users may also plan the drone's route ahead of time, and the device can use computer vision to take photos that can be saved for later use. SkySquirrel use algorithms to integrate and assess the photos and data collected. As a result, a comprehensive assessment of the vineyard's health will be produced. After a successful joint venture, SkySquirrel Technologies and Vine View Scientific Aerial Imaging Inc. merged in January 2018.

**FarmShots:** Satellites for agricultural health and sustainability monitoring are their specialty. FarmShots is a company that specializes in analyzing agricultural data from satellite and drone images. The corporation's headquarters are in Raleigh, North Carolina. FarmShots is a project that tries to diagnose illnesses, pests and insufficient plant nourishment on farms.

**Abundant Robotics:** One of their specializations is apple harvesting technology. With the goal of providing robotic equipment for the most demanding agricultural jobs, the Abundant Robotics team has broken new ground in a range of disciplines. Despite the fact that apple harvesting is the company's core emphasis, it has spent the previous two years focusing on complex agricultural issues and developing products based on their successful research. Abundant Robotics' first product was an apple picking machine.

**Ibex Automation:** One of their specialties is agricultural robotic systems. The company, based in Wortley, United Kingdom, develops self-driving agricultural robots for farmers. The Ibex Automation system includes a grassland-specific autonomous precision weed detection and spraying system in addition to autonomous precision weed detection and spraying. Ibex Automation Ltd was created as a result of an Innovate UK-funded prototype project.

**Beriqo:** Beriqo, a Canadian startup, is developing a system for managing agricultural data. To detect pixel-by-pixel changes in weekly satellite image stacks, the company employs cutting-edge deep learning techniques. Their research supports precision agriculture decision-making by identifying criteria that have previously been neglected, such as underperforming fields, nutrient deficiency, pest damage and fertilizer stripes.

**AgEYE Technologies:** AgEYE Technologies, situated in the United States, creates intelligent software and sensor systems to improve agriculture's quality, predictability and profitability. They employ a combination of machine vision, artificial intelligence and machine learning to detect diseases and pollution in all of the plants far more quickly than people can. Their techniques dramatically minimize the amount of time and labor required to scour the farm for infections and anomalies.

## 7.8 INDIAN START-UP

Agriculture is acknowledged as the economic underpinning, as well as a substantial revenue and product-generating business. The globalization of technology has also delved into this industry, resulting in more efficient outcomes. Farmers can gain a better understanding of agricultural yield by using AI, automation and robotics.

**AgriTech:** AgriTech has flourished in India, with a spate of companies leveraging data analytics, machine learning and satellite images, among other technologies, to assist farmers in increasing their output. AgriTech enterprises are favored by the Indian Government's Startup India program, according to NASSCOM research. Over 350 AgriTech companies raised $300 million in funding worldwide in 2016, with India accounting for 10% of the total.

**SatSure:** Since its beginning in early 2016, SatSure has been at the forefront of bringing the best practices of satellite image processing, big data capabilities, and IT to agriculture. It also aspires to help farmers better their lives by developing intervention and decision intelligence frameworks for agri-value chain partners, as well as assisting with crop insurance, agro loan service innovation, and market connections. The firm has created mobile app platforms that provide data on crop supply and crop stressing in their area. This helps farmers decide what to plant, when to water or fertilize, and when to harvest.

**Fasal:** Fasal's microclimate forecasts are tailored to each farm and are calculated on a point scale rather than on a kilometer-wide spatial scale. According to Fasal's founder, the AI-based microclimate forecasting system incorporates real-time in-field data and links it to publically available weather forecasts, allowing farmers to profit from real-time, actionable information relevant to day-to-day agricultural operations.

**Aibono:** This organization, India's first smart agricultural collective, employs the internet, artificial intelligence and shared services to help small farmers improve their fortunes. Precision agriculture technology underpinned by real-time supply and demand synchronization are available through the Agri 4.0 collective. One of its solutions is real-time precision agriculture as a service to farmers. The company presently employs about 60 employees, with plans to expand to 100 by the end of 2018.

**Gobasco:** This Gurgaon and Lucknow-based AgriTech company leverages real-time data analytics on data-streams from various sources across the country, backed by AI-optimized automated pipelines, to improve and boost the efficiency of the agricultural supply chain. Both suppliers and customers benefit from the startup's data-driven online agri-marketplace, which puts the best prices in their hands. The company's products include transaction discovery, procurement optimization and real-time transportation optimization, to name a few. Last year, Matrix Partners gave the startup an undisclosed sum of initial money.

**Intello Labs:** This Bengaluru-based company has created computer vision-based solutions that use images as critical data to generate insights and actionable recommendations, earning it the title of India's most honoured AgriTech company. Crop inspection and agricultural product grading are handled by Intello Labs' two main agricultural products. Both products read photos and provide quality parameters based on the input data. The venture combines emerging technologies like deep learning, AI and IoT to help farmers scale their businesses effectively.

## 7.9 SOME USEFUL EXAMPLES ASSOCIATED WITH THE AGRICULTURE SECTOR

Machine learning is an application of AI that allows a system to learn and improve without having to be explicitly programmed. Machine learning is concerned with the creation of computer programs that can access data and learn on their own. The first example of soil analysis and prediction of suitable crop proposed by [30] are outlined below.

### 7.9.1 Soil Analysis and Prediction of Suitable Crop

Most agriculture is not working in India due to lack of knowledge about agriculture and fertilizer use. Currently soil ingredients are only tested in the soil testing laboratory, where they use the old method. In an existing system the soil can only be checked for fertility and moisture level. It should be given a soil testing lab. It will take some days to get the result. Farmers are struggling to get farm survey reports quickly. Farmers fear rain every year for their demand yield. In the classical way, various soil elements can be suggested with the help of pH. As soon as nutrients are available, recommendations for planting a specific crop will be provided with the help of a pH electrode

A. Preparation of Soil Analysis Samples in Existing System

- Handled in the laboratory

- Sample drying

- Care after drying

B. Limit of Existing Program

- The recovery period will take 15 to 20 days.

- One soil sample cannot predict global performance.

- The farmer does not have enough information to know which fertilizer to use.

- The chemical methods used in the laboratory are very dangerous.

- Managing many soil samples from the lab is impossible for farmers.

## 7.9.2 Proposed System

Our program will overcome the constraints of the old methods used to date. Here soil parameters and nutrients are taken as NPK to determine the fertility level of that soil. Along with soil analysis our system will also predict plants using machine learning. It compares the current data with the available data collected from the Department of Horticulture and Agriculture in different categories such as pH, EC, humidity, and temperature values. Farmers can inspect the soil many times during the cultivation process and take precautionary measures to get the best yield. At the end, reports will be produced so that farmers can keep track of their offspring.

Sensors Used:

- pH sensor
- Soil moisture sensor
- Soil temperature sensor

### 7.9.3 Use of Machine Learning

Machine learning is the use of AI that gives the system the ability to automatically read and develop from information without being explicitly programmed. Machine learning focuses on the development of computer systems that can access the data and use it for their own learning.

1. Algorithm Used for this Project: Supervised learning algorithm. The algorithm assists in making predictions about the data in the training program and detects the process optimization itself. There is only a learning curve once the algorithm has achieved an acceptable rating or performance level.

2. Types of problem which is a suitable particular crop and agriculture scenario.

3. Supervised Learning: In supervised learning, we begin by introducing a database containing training attributes and target characteristics. The supervised learning algorithm will study the relationship between training examples and their intended flexibility and use learned relationships to distinguish new entries (without enemies).

4. The assistance: The accuracy of a classifier is measured by the percentage of test tuples that the classifier correctly classifies. The easiest approach to examine the different algorithms and attempt different settings within each algorithm is to have access to high accuracy. Cross validation can be used to choose the best. Different algorithms must take into account criteria such as accuracy, training time, compliance, the number of parameters, and special charges when choosing the optimal algorithm for the job.

### 7.9.4 Data Sets

Measurement data using sensors and dataset collected from the Department of Horticulture and Agriculture are both embedded in a machine learning python. The algorithm we used here will compare both values, there will be a straightforward match. Corresponding data will appear on the screen so that we can predict the optimal yield of a particular soil and, based on nutrient status, can increase the necessary soil fertilizers that play an important role in the development of farmers' crops.

### 7.9.5 Fertilizer Recommendation

There are two ways to make a fertilizer recommendation:

- A recommendation that reflects the nutritional and yield requirements of a high-yielding economy based on one or more soil moisture conditions.

- "Target yield recommendation" that reflects the nutritional requirements of a variety of fertilizers at lower and higher yields under similar humidity conditions. With this information, producers have the flexibility to choose a fertilizer application or target yield that best fits their situation.

## 7.10 CONCLUSION

Soil analysis and forecasting of a suitable crop provides crop and fertilizer recommendations based on the use of different soil components. Farmers can get soil testing equipment at the door. This project replaces the first soil test effectively and therefore farmers know immediately about their soil. The result provided by this project helps the farmer to make a decision and prevents them from using uneven fertilizer. This project is very useful for second-level farmers.

In paper [31] the authors explore another topic related to "**Random Forest Algorithm for Soil Fertility Prediction and Grading Using Machine Learning**". The project is a vision to place the soil in a specific location and raise the world's best yield based on previously fed data in parallel. They proposed their first model based on soil formation rate dependent on many organic compounds, and micro-nutrients are an important methods adopted by the models. The second model is based on soil type and nutrient content, and a classification algorithm will be applied to plant crops.

**System Design:** Module1–Soil Formation. This module includes a machine learning model designed to help the farmer understand tehsoil quality by looking at various soil factors, and based on these criteria the following model is developed. Module 2: Crop Recommendation – This study has the final design to help farmers increase their yields and make better use of farmland, taking into account the minimum nutrient requirements and soil types needed to grow a particular crop.

**Data Processing:** Soil properties such as PH, EC, OC, S, K, Zn, Mn, B and soil types are considered to be variation factors, and non-significant values in the dataset are eliminated. The data collected will be processed using appropriate methods of transforming agriculture, by improving crop production and, so, improving the livelihood of farmers. The project helps restore soil by utilizing soil health structure as a threshold to release capacity of rain-based farming areas. Any machine learning project requires data processing.

**In order to find the best model fit for the data, you'll need to do the following:**

The usage of classification techniques such as support vector machine, random forest classification, and decision tree is made, and the right model is chosen based on the RMSE (root mean squared error). After gathering all of the data, we must choose the correct column of required data and eliminate the remainder.

## 7.11 ALGORITHMS

MLVR cost function: provides the best fit line (minimize mean squared error).

$$\text{Minimize}\left(\frac{1}{n}\right)\sum_{i=1}^{n}\left(pred[i]-y[i]\right)^2$$

Gradient Descent: An algorithm that reduces the cost function and gets the best fit line.

**Data Model Training and Testing:** Figure 7.8 show the flow charts of data model training and testing proposed by [32, 33].

When the goal value is continuous, regression is a statistical method for making a forecast (using linear regression).Simple linear regression is a single independent variable. Multi variate linear regression is a type of linear regression that uses many independent variables. On a scale of 1 to 5, we use the multi-variate linear regression algorithm to forecast soil fertility.

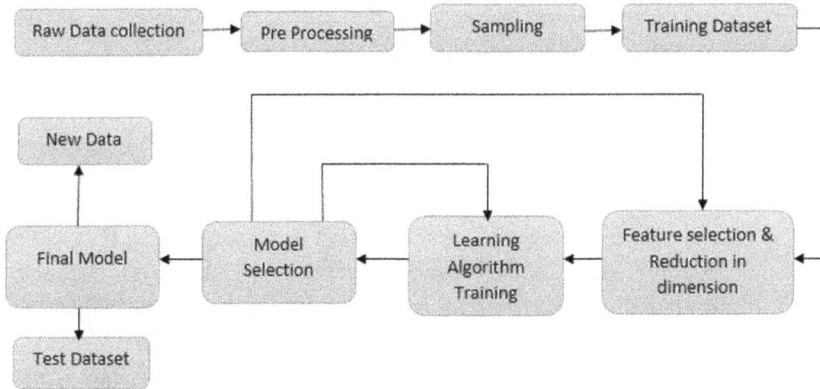

FIGURE 7.8    Flow charts of data model training and testing [33].

**Random Forest Classifier:** Random forest is a supervised learning algorithm that creates multiple decision trees and can be pooled together to get an accurate and consistent forecast. This allows us to add randomness to our model. When dividing any node, random forest searches for the most important parameter of all, and then finds the best among them by the height of random symbols. This ultimately creates a model with the highest accuracy in various fields. Only the properties selected in this algorithm are taken by the location partition [34–37]. Trees can be further randomized, using random entry for a defined element instead of searching for the best possible entry. The algorithm is stochastic forest training that implements the simple process of implementing aggregation or bagging for tree learners. To reduce our model variability, we used the boot process without increasing the bias, which ultimately leads to model efficiency. By training multiple trees in a dataset, we can generate fully correlated trees, and by resetting these trees we call bootstrap models [38–41].

**Result Analysis and Conclusion:** Although ground-line is based on their properties, nonlinear regression proves to be an efficient algorithm with very little error, that is, a metric for measuring accuracy when there is a problem with regression. In terms of crop forecasting, random forest shows that Gaussian is a better team than Naive Bayes and support vector machine. This model helps to estimate the soil fertility class as a decision algorithm using a variety of algorithms where large amounts of data are available. The system employs machine-based algorithms such as multi-variable linear regression, support vector machine, and random forest classification to produce good fault analysis findings. These produce the best and most accurate results. As a result, the project aids in the reduction of farmer conflict.

## REFERENCES

1. Yoon, C., Huh, M., Kang, S.-G., Park, J., & Lee, C. (2018). *Implement smart Farm with IoT Technology*. Paper presented at the 2018 20th International Conference on Advanced Communication Technology (ICACT).
2. Shahzadi, R., Tausif, M., Ferzund, J., & Suryani, M. A. (2016). Internet of things based expert system for smart agriculture. *International Journal of Advanced Computer Science and Applications*, 7(9), 341–350.
3. Das, S. D., Deb, N., Biswal, G. R., & Das, S. (2019). *High Voltage Aspects of Smart Agriculture through GIS Towards Smarter IoT*. Paper presented at the 2019 International Conference on Automation, Computational and Technology Management (ICACTM).
4. Patel, G.S., Rai, A., Narayan Das, N., & Singh, R.P. (Eds.). (2021). *Smart Agriculture: Emerging Pedagogies of Deep Learning, Machine Learning and Internet of Things*, CRC Press. https://doi.org/10.1201/b22627.
5. Bhagat, M., Kumar, D., & Kumar, D. (2019). *Role of Internet of Things (IoT) in Smart Farming: A Brief Survey*. Paper presented at the 2019 Devices for Integrated Circuit (DevIC).
6. Amrita Rai, Deepti Sharma, Shubhyansh Rai, Amandeep Singh, Krishna Kant Singh (2021). "IoT-Aided Robotics Development and Applications with A'". Emergence of Cyber Physical System and IoT in Smart Automation and Robotics.
7. Alreshidi, E. (2019). Smart Sustainable Agriculture (SSA) Solution Underpinned by Internet of Things (IoT) and Artificial Intelligence (AI). *arXiv preprint arXiv:1906.03106*.
8. Citoni, B., Fioranelli, F., Imran, M. A., & Abbasi, Q. H. (2019). Internet of Things and LoRaWAN-Enabled Future Smart Farming. *IEEE Internet of Things Magazine*, 2(4), 14–19.
9. Zhu, N., Liu, X., Liu, Z., Hu, K., Wang, Y., Tan, J., . . . Jiang, Y. (2018). Deep learning for smart agriculture: Concepts, tools, applications, and opportunities. *International Journal of Agricultural and Biological Engineering*, 11(4), 32–44.
10. Shi, X., An, X., Zhao, Q., Liu, H., Xia, L., Sun, X., & Guo, Y. (2019). State-of-the-art internet of things in protected agriculture. *Sensors*, 19(8), 1833.
11. Yazici, M. T., Basurra, S., & Gaber, M. M. (2018). Edge machine learning: Enabling smart internet of things applications. *Big Data and Cognitive Computing*, 2(3), 26.
12. Varman, S. A. M., Baskaran, A. R., Aravindh, S., & Prabhu, E. (2017). *Deep Learning and IoT for Smart Agriculture using WSN*. Paper presented at the

2017 IEEE International Conference on Computational Intelligence and Computing Research (ICCIC).

13. Valecce, G., Strazzella, S., Radesca, A., & Grieco, L. A. (2019). *Solarfertigation: Internet of Things Architecture for Smart Agriculture.* Paper presented at the 2019 IEEE International Conference on Communications Workshops (ICC Workshops).

14. Byabazaire, J., Olariu, C., Taneja, M., & Davy, A. (2019). *Lameness Detection as a Service: Application of Machine Learning to an Internet of Cattle.* Paper presented at the 2019 16th IEEE Annual Consumer Communications & Networking Conference (CCNC).

15. Goap A., Sharma, D., Shukla, A., & Krishna, C. R. (2018). An IoT based smart irrigation management system using Machine learning and open source technologies. *Computers and Electronics in Agriculture*, 155, 41–49.

16. Bu, F., & Wang, X. (2019). A smart agriculture IoT system based on deep reinforcement learning. *Future Generation Computer Systems*, 99, 500–507.

17. Kumar, A., Goyal, S., & Varma, M. (2017). *Resource-Efficient Machine Learning in 2 KB RAM for the Internet Of Things.* Paper presented at the Proceedings of the 34th International Conference on Machine Learning-Volume 70.

18. Dharmasena, T., de Silva, R., Abhayasingha, N., & Abeygunawardhana, P. (2019). *Autonomous Cloud Robotic System for Smart Agriculture.* Paper presented at the 2019 Moratuwa Engineering Research Conference (MERCon).

19. Jin, X.-B., Yang, N.-X., Wang, X.-Y., Bai, Y.-T., Su, T.-L., & Kong, J.-L. (2020). Hybrid deep learning predictor for smart agriculture sensing based on empirical mode decomposition and gated recurrent unit group model. *Sensors*, 20(5), 1334.

20. Elijah, O., Rahman, T. A., Orikumhi, I., Leow, C. Y., & Hindia, M. N. (2018). An overview of Internet of Things (IoT) and data analytics in agriculture: Benefits and challenges. IEEE Internet of Things Journal, 5(5), 3758–3773.

21. Jiang, W., Wang, Y., & Qi, J. (2019). *Study on the Integrated Model of Modern Agricultural Variety Breeding in the Internet of Things Environment.* Paper presented at the Proceedings of the 2019 Annual Meeting on Management Engineering.

22. Katyal, N., & Pandian, B. J. (2020). A Comparative Study of Conventional and Smart Farming *Emerging Technologies for Agriculture and Environment* (pp. 1–8): Springer.

23. Kitpo, N., Kugai, Y., Inoue, M., Yokemura, T., & Satomura, S. (2019). *Internet of Things for Greenhouse Monitoring System Using Deep*

*Learning and Bot Notification Services.* Paper presented at the 2019 IEEE International Conference on Consumer Electronics (ICCE).

24. Khanna, A., & Kaur, S. (2019). Evolution of Internet of Things (IoT) and its significant impact in the field of Precision Agriculture. *Computers and Electronics in Agriculture*, 157, 218–231.

25. Mohapatra, S. K., & Mohanty, M. N. (2019). Analysis of Diabetes for Indian Ladies Using Deep Neural Network *Cognitive Informatics and Soft Computing* (pp. 267–279): Springer.

26. Zhang, M., Changying, L. & Fuzeng, Y. (2017). Classification of foreign matter embedded inside cotton lint using short wave infrared (SWIR) hyperspectral transmittance imaging. *Computers and Electronics in Agriculture*, 139, 75–90.

27. Kumar, A., Goyal, S., & Varma, M. (2017). *Resource-Efficient Machine Learning in 2 KB RAM for The Internet of Things.* Paper presented at the Proceedings of the 34th International Conference on Machine Learning-Volume 70.

28. Veloo, K., Kojima, H., Takata, S., Nakamura, M., & Nakajo, H. (2019). *Interactive Cultivation System for the Future IoT-Based Agriculture.* Paper presented at the 2019 Seventh International Symposium on Computing and Networking Workshops (CANDARW).

29. Sharma, H., Haque, A., & Jaffery, Z. A. (2019). Maximization of wireless sensor network lifetime using solar energy harvesting for smart agriculture monitoring. *Ad Hoc Networks*, 94, 101966.

30. Panchamurthi, S. (2019). Soil Analysis and Prediction of Suitable Crop for Agriculture using Machine Learning. International Journal for Research in Applied Science and Engineering Technology, 7, 2328–2335. 10.22214/ijraset.2019.3427.

31. Fuentes, A., Yoon, S., Youngki, H., Lee, Y., & Park, D. (2016). *Characteristics of Tomato Plant Diseases: A Study for Tomato Plant Disease Identification.* Paper presented at the Proc. Int. Symp. Inf. Technol. Converg.

32. HHaykin, S. (2010). *Neural Networks and Learning Machines*, 3/E: Pearson Education India.

33. Keerthan Kumar T. G., Shubha C., Sushma S. A. (2019) Random Forest Algorithm for Soil Fertility Prediction and Grading Using Machine Learning, International Journal of Innovative Technology and Exploring Engineering (IJITEE) ISSN: 2278-3075, 9(1).

34. K. Rusia, S. Rai, A. Rai & S. V.Kumar Karatangi (2021), "Artificial Intelligence and Robotics: Impact & Open issues of automation in Workplace", 2021 International Conference on Advance Computing and Innovative Technologies in Engineering (ICACITE), 2021, pp. 54–59, doi: 10.1109/ICACITE51222.2021.9404749

35. Ale, L., Sheta, A., Li, L., Wang, Y., & Zhang, N. (2019). *Deep Learning Based Plant Disease Detection for Smart Agriculture.* Paper presented at the 2019 IEEE Globecom Workshops (GC Wkshps).

36. Khanna, A., & Kaur, S. (2019). Evolution of Internet of Things (IoT) and its significant impact in the field of Precision Agriculture. *Computers and electronics in agriculture*, 157, 218–231.

37. Khoa, T. A., Man, M. M., Nguyen, T.-Y., Nguyen, V., & Nam, N. H. (2019). Smart Agriculture Using IoT Multi-Sensors: A Novel Watering Management System. *Journal of Sensor and Actuator Networks*, 8(3), 45.

38. Mohapatra, S. K., & Mohanty, M. N. (2020). Big Data Analysis and Classification of Biomedical Signal Using Random Forest Algorithm. *New Paradigm in Decision Science and Management*, (pp. 217–224): Springer.

39. Ramdinthara, I. Z., & Bala, P. S. (2019). *A Comparative study of IoT Technology in Precision Agriculture.* Paper presented at the 2019 IEEE International Conference on System, Computation, Automation and Networking (ICSCAN).

40. Krishnan, R. S., Julie, E. G., Robinson, Y. H., Raja, S., Kumar, R., & Thong, P. H. (2020). Fuzzy Logic based Smart Irrigation System using Internet of Things. *Journal of Cleaner Production*, 252, 119902.

41. Miss. Snehal, S. D. (2014). Agricultural Crop Yield Prediction Using Artificial. International Journal of Innovative Research in Electrical, Electronic.

# CHAPTER 8

# Deep Learning in Smart Agriculture Applications

Amrita Rai[1] and Asha Rani Mishra[2]

[1]*SMIEEE & Associate Professor, GL Bajaj Institute of Technology and Management Greater Noida, India*
[2]*GL Bajaj Institute of Technology and Management Greater Noida, India*
*E-mail: amritaskrai@gmail.com; asha.mishra@glbitm.ac.in*

## 8.1 INTRODUCTION: BACKGROUND AND MOTIVATION

Agriculture is the backbone of our economy, where, thanks to increased populations, the need for foodstuff is consistently rising. There's a huge need for advancements in the agriculture sector to meet these requirements, by using precise calculations regarding yield production, and by using the best and latest farming equipment, and so on, with reference to digital farming, precise farming, intelligent farming, intensive farming, continual farming, organic farming and agribusiness, which come under the umbrella of 'fashionable farming' [1]. Firstly, precision agriculture may be a farming supervision theory counting on detecting, computing and reacting to inconsistency within an equivalent field and other field yields. The main objective of precision agriculture study is constructing a judgment support system for managing the whole field farming with the aim of optimizing profits on inputs alongwith the preservation of resources. Predicting weather patterns, and using different fertilizers, remote sensing and sensors for crop health, are the initial step for precision farming [2].

Secondly, 'agribusiness' is that the professional term associated with agricultural yields. It's a hybrid of business in agriculture that consists of

DOI: 10.1201/9781003217237-8

**165**

breeding, yield production, agrichemicals, farm appliances and seed supply, and also the strategy forselling and distribution. The representatives, and therefore the organizations, that affect the food and fiber chain are a part of this agribusiness structure [3].Another important part of modern agriculture is dealing with environmental issues such as yield, environmental effect, food safety and sustainability in the current environment. Because worldwide output is fast expanding, a massive increase in crop production is required, as well as consistent availability and good nutritional quality. This will be achieved by protecting the natural ecosystem by using sustainable farming methods. Thee farming management concept supports observing, measuring and responding to inter and intra-field variability in crops [4].

Aside from these characteristics of the contemporary agriculture sector, the area of agriculture faces a variety of issues, such as improper farm treatment, numerous animal ailments, pest infestation and poor irrigation, to name a few. All of these issues wreak havoc on crop production, while also posing a risk to the ecosystem due to the excessive use of chemicals.It is impossible to find a universal solutionto all of the problems in order to overcome this dilemma. The composite, intermittent agricultural ecology can be treated with instantaneous observation and inquiry in all aspects and occurrences in order to report these difficulties. Artificial intelligence in general, and deep learning in particular, can assist to alleviate this problem.Agriculturalists will be able to increase crop yield while lowering beginning costs and balancing losses caused by natural disasters using deep learning techniques.

## 8.2 POPULAR DEEP LEARNING ARCHITECTURES USED IN THE AGRICULTURAL DOMAIN

Deep learning is a cutting-edge technique that has found success in a variety of fields. It has a wide range of applications, including image processing and text classification. Because deep learning has a high success rate in other domains, it has been applied to agricultural operations. Deep learning is made up of multiple layers of neural networks, which are designed to solve complex tasks. Some of the deep learning models provide remarkable results; in terms of scale they are unmatched by humans. Each layer uses the result of previous results as input, and the whole network is trained as one chain.A deep learning platform is a tool that allows users to create various deep learning architectures or to apply deep learning to

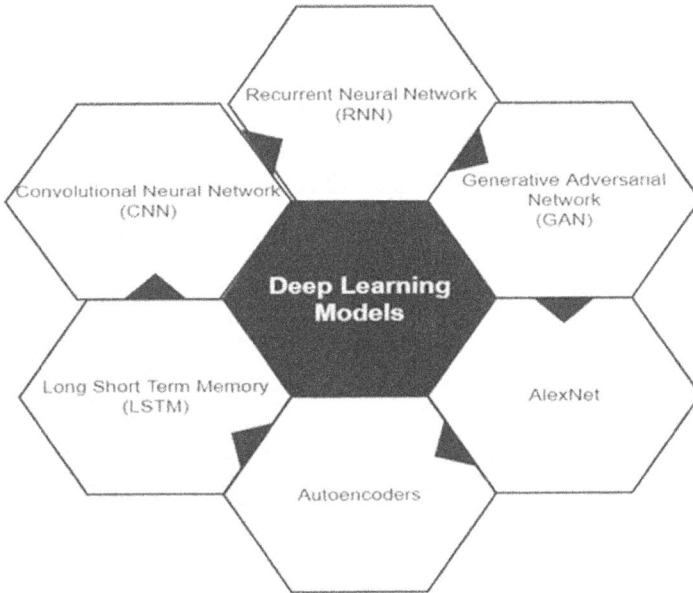

FIGURE 8.1    Popular deep learning architectures used in the agricultural domain.

a variety of commercial applications through apps and services, as shown in Figure 8.1.

One of the primary differences between machine learning and deep learning is that deep learning requires more data for classification, but machine learning requires only a limited amount of data for categorization [5]. The most popular deep learning tools are theno, kera, tensor-flow, py-torch and tool-box. Examples of deep learning are image processing and text classification. Some of the deep learning architects are as follows.

### 8.2.1  Convolutional Neural Networks

Convolutional neural networks (CNNs) are based on a feed-forward neural network and is designed on an animal cortex, using multi-layer perceptron for this process. In CNNs, the minimum amount of pre-processing, recti-fying the linear unit's activation function, is often used. General applications are image/video recognition, natural language processing, chess and so on. Convolution is used to find the features which are similar by using different places of images. It is conducted using learnable filters, which are passed through the input data/images. An example is depicted in Figure 8.2 as CNN in plant disease detection.

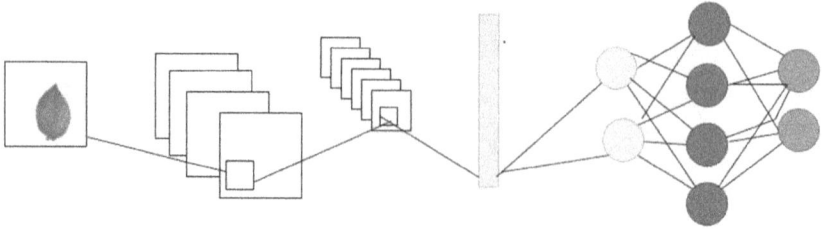

FIGURE 8.2    CNN in plant disease detection.

Source: https://medium.com/bbm406f19/week-2-plant-disease-detection-9bdd819b870)

The technique used to increase the data set and improve CNN accuracy is known as data augmentation. Provided there is alarge enoughbig dataset, CNN increases the accuaracyof correct classification. Some of the applications of artificial neural networksare decoding facial reorganization, document analyzing, historic and environment collection, understanding of climate and advertising.

## 8.2.2 Recurrent Neural Networks

Recurrent neural networks (RNNs) area type of neural network where output of the previous loop is considered as input for the current loop. General applications of generative neural networks are speech recognition, hand written recognition and analysis of sequence of data. Also generative neural networks automatically generate programming codes that give a pre-defined objective. The working process of RNN consists of providing input to the model. Representation of the data in the input layer is computed and sent to the hidden layer, where it conducts sequence modeling and training in a forward or backward directions. Multiple hidden layers can also be used, final hidden layer sends the processed result to the output layer. Long-short-term memory RNN is currently a popular RNN model. It is effective on data sequences that require memory or details of the last events. Some of the applications of RNN are language modeling and prediction, speech recognition, machine translation, image recognition and translation. Long-short-term memory is the latest improvement of RNN networks; these networks are known as cells [6–10]. These cells consider the input from previous state and present input, and also decide which information needs to be considered, and which should be disregarded. The previous condition, present memory and the current input combine together to predict the next output. Figure 8.3 shows the RNN in crop classification.

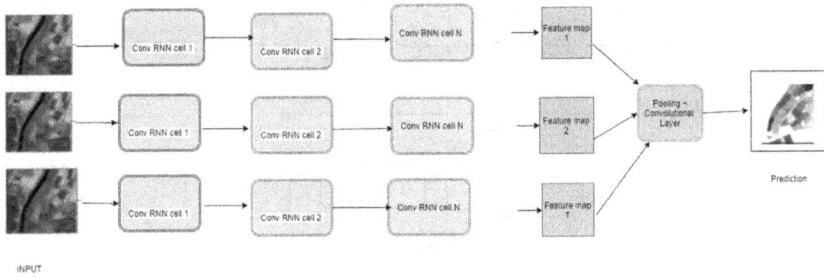

FIGURE 8.3   RNN in crop classification.

Source: https://prs.igp.ethz.ch/research/current_projects/large-scale-crop-classification-from-satellite-images.html

## 8.2.3 Generative Adversarial Networks

The novelty of generative adversarial networks (GANs) lies in the technicality of their design. It is a type of unsupervised machine learning which includes computerized innovation to understand the similarities or prototype data in the manner that system produces the resultant. GANs are smart models that build a productive system by modeling a problem, having two sub-models as a part of supervised learning. They are generative systems that can be educated to create illustration. GANs are the latest and most innovatve technology. It works due to the potential of generative practical model in their ability. The system is related to picture-to-picture conversion, as in [10–15]–converting pictures of one season to another (e.g. summer scenes to winter scenes) or day images to night ones, in creating photo-realistic images of items, scenes and individuals, which the individuals recognize as forged.

## 8.3 APPLICATION OF DEEP LEARNING IN AGRICULTURE

Deep learning has transformed the agriculture sector to a new level, using CNN, RNN and GAN techniques. This provides better results and encourages the agriculture domain to grow with advanced technology. The procedure of deep learning uses a current procedure in processing for image and studying the knowledge with efficient result and great potential. The intense growth in the field of deep learning has shown many general leads, but is emerging as a boon in agriculture especially. By comparing deep learning with one prevailing common procedure, it is often said this that deep learning is outperforming existing commonly used image processing techniques and has greater precision [16]. Deep leaning allows

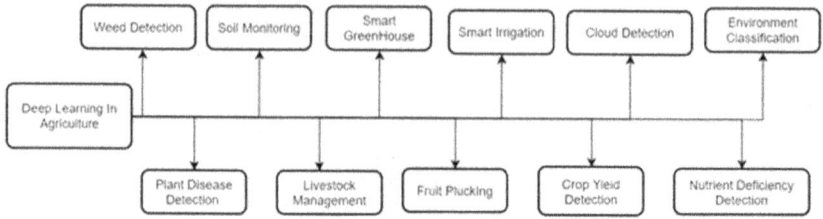

FIGURE 8.4  Applications of deep learning in agriculture.

mathematical systems which are made from many processing stages to represent information with many level of abstraction. Neural networks and back propagation are the icore dea of deep learning. Figure 8.4 describes the different areas of agriculture where deep learning is utilized for various purposes.

Deep learning models' highly hierarchical structure and vast learning capacity enable them to perform particularly well in classification and prediction, as well as being adaptive to a wide deep learning is most commonly associated with raster-based data (e.g., video and images), it may be used with any type of data, including audio, speech, and natural language, as well as continuous or point data such as meteorological data [18], soil chemistry [19] and population data [20].

## 8.4  CNNS IN AGRICULTURE RNNS

Because of its high capacity for image processing, the CNN is often employed in agriculture. Plant or crop classification, pest and yield prediction, robot harvesting, and disaster monitoring are some of the major applications of deep learning in agriculture. Leaf image and pattern classification can be used to create a disease recognition model for plants. The Berkeley Vision and Learning Center has created a new deep learning framework for developing a disease detection model. This technique, in addition to detecting 10–15 cases of disease from healthy leaves, can also isolate plant leaves from their surroundings. In 2007, a technique that combines CNN and K-mean features was developed to control and identify weeds. [21] gives an example of a CNN, combining convolutional and fully connected (dense) layers.

For yield prediction and robot harvesting, fruit counting is one of several important factors; we cannot produce satisfactory results through traditional counting or video or camera image counting, and also these processes are time consuming. Pre-processing of those sorts of images is challenging due to occlusion and illumination. [22] has introduced the technique to spot the livestock animal like pig face recognition through the CNN technique. Conventionally, frequency identification tags were used for detecting the animals, which was a cumbersome job. To accompany a fully convolutional network, a method known as blob detection is proposed. The first step is to gather the human formed labels from the set of fruit images and then this model is trained for an image segmentation performance. Then CNN was trained to consider the bifurcated pictures and generate a middle approximation of the fruit count. The concluding stage of the work is to train a regression equation to map intermediate fruit count estimation to final human generated label count. Accuracy as well as efficiency is increased by combining deep learning with blob detection.

Land classification classifies the land use and covers disaster assessment of risk, food and agriculture. The overall idea of the deep learning method is to integrate information developed by multiple heterogeneous sources using machibe learning techniques to provide information processing and picturing capability. This process has four steps: 1. Noise filtration and data clustering; 2. Clarification of land cover; 3. Map post processing; 4. Geospatial analysis.

For high quality image processing, unmanned aerial vehicle can be used. The reorganization of items is emerging in the agriculture sector. Especially for farmers, the finding of obstacles is important, and this is increased by the use of of highly autonomous machines. It is not easy to operate such machines without supervision. So detection of real-time risk automatically with high reliability is more important. For sustainable land use certain conditions need to be considered: planning of reducing $CO_2$ emission; diminishing land degradation; and improving economic returns using valuable information, which we get from satellites. For decision making in precision agriculture and agro industry, CNN and genetic algorithm can be convenient methods by using translated satellite images [19]. Weather forecasting is one of the main factors for farmers that can be predicted using CNN. Crop yield estimation is also a key factor for farmers, consumers and for the government to predict the crop harvest. CNN is not only used for only crop estimation or agriculture purpose but also used in classifying animal behavior [23–24].

## 8.5 DEEP RNNS IN AGRICULTURE GANS

Land cover classification isa key challenging area in agriculture, the main point beingto recognize the type of land and quality of that land. In the past, many applications are ignored time series effect and were based on mono-temporal observation. Mono-temporal methods can be dependent on factors such as weather.

To solve the problems related to RNN, the model known as NARX is introduced, where NARX means nonlinear autoregressive model process with exogenous input. Previous prediction values are considered as input and present and previous values of exogenous input. The system not only judges the independent inputs but also the previous response of the system, which makes itmore powerful. By considering the NARX model, another model is developed, which is NARXNN, for estimation of time series of Leaf Area Index.

RNN is also used to predict the weather. [1,25] designed three models for prediction of weather: NARXNET, CBR (case-based reasoning model) and SCBR (segment case-based reasoning model). Other authors [1] also produced a model which is used to identify the location of animals in the forest. For presenting this model theyused particle swarm optimization algorithm combined with an RNN model; the results obtained by this model has less errors.The genotype of maize and rice was mapped and trained with a deep neural network (DNN) using the crop. It was entirely based on categorization prediction, with a 64% estimation accuracy. The atmospheric parameters are compared and mapped with the sensor and drone photos in the following type of prediction utilizing regression method.

## 8.6 GANS IN AGRICULTURE

The GAN is considered as one of the most useful neural networks in many fields. Mainly GAN is used to find the feature loss in image processing caused by down sampling. When the image is reduced or condensed, some of the information may be lost or the quality of that will be image will be lost, so we may need to recover all the original details from it. For that, there is a perpetual loss function comprised of an adversial loss and content loss. This function is compared with widely used pixel-wise Mean Square Error loss. When the model is trained with a greater number of images, this model would be able to improve highly compressed images. This has high importance value since this is associated with all models which contain image processing work, mainly in agriculture, because certain applications are dependent on remote sensing images.

Barth et al. [6,23] proposed amodel to overcome problems associated with the largeamount of data in deep learning systems. The shortage of manually marked information and large quantity of data (DL model, GAN based model), is used. This is also called an unsupervised cycle, generative adversarial system, to optimize the practicality of artificial agricultural pictures. The authors have proposed 10500 artificial, 50 empirically annotated and 225 unlabeled empirical pictures to get their model trained. The hypothesis was that there was resemblance between synthetic images and empirical images, which can be enhanced qualitatively to improve the transformation of features. This analysis resulted inthe artificial pictures beingtransformed easily on local characteristics like light diffusion, color and consistency as compared to global feature translation, which was not that good.

## 8.7 CHALLENGES IN AGRICULTURE

Generally, in machine learning, it is not easy to analyzedata which is not structured. For this type of data analysis, applying deep learning methods will be more useful, where we can use different types of data format to train the algorithm in deep learning. To find the relation between any different domains which are interdisciplinary we can use the deep learning algorithm. Generally, workers get tired or irresponsible or neglect small things, but in deep learning models that's not the case. They will perform thousands of cycles of work without any error, and in a short period of time. Also, the quality of work will not reduce until and unless data given by the user has some problem. In traditional learning approaches, identification of features need to be accurate, where as in deep learning, models have the ability to create new features by themselves. Generally, problem solving in machine learning is done by dividing tasks into small tasks and combining the results of all the small tasks for the final output, whereas in deep learning, tasks are solved on an end to end basis. Deep learning requires large amounts of data or information; if the information is less, we will not get the exact output. It is expensive to train a deep learning model. One of the major disadvantages is we are not able to find how the analysis will take place inside the model. Generally, we call it as black-box; it is important because interpretability is necessary in some domains. Nowadays as machine learning is growing in all domains in a dramatic way, the main fear is that machine learning may take over all the work done by humans, driving humans into unemployment or slavery.

## 8.8 SOME USEFUL EXAMPLES ASSOCIATED WITH THE AGRICULTURE SECTOR

**Soil Fertility Grading:** This is an essential issue in agriculture proposed by [9, 25], and various complex methods have recently been applied, including Bayesian networks, neural networks and D-S theory. When a valid model can be simulated, most present approaches assume that a large enough dataset is available. However, in the context of soil preparation, training and collecting data from soil testing is an expensive and time-consuming operation that is sometimes impossible.

**Bayesian Network Learning:** There are two pieces to a Bayesian Network (BN). The joint probability distribution of a set of random variables X=(X 1,X 2,...X n) is represented by BNs G (qualitative and quantitative parts):

G is a directed acyclic graph (DAG), in which each node corresponds to a random variable in X, and G records the probability distribution's (condition) independencies.

$\theta$ is a conditional probability table, abbreviated as CPT, which encodes each family's conditional distributions (a node and its parent node), = {p(X i | $\pi$ Xi)|1 $\leq$ i $\leq$n} ($\pi$ Xi is the parent nodes of Xi).

Structure learning and parameter learning are both part of the BN's learning algorithm. There are two methods:

The task is an optimization issue; hence a score-based technique is used. The aim is to solve a constraint fulfilment problem using a constraint-based strategy.

Let's use a set of discrete random variables to represent the soil fertility grading task for number k land squares.

Let $X_k = (X_1, X_2, ..., X_n)$, there are n land squares, (k=1..n), training data sets are $D^k$, (k=1..., n).

Except for the target land square (1 $\leq$ t $\leq$ n), which has a very tiny training data set, the remainder of the squares, referred to as auxiliary squares, contain enough data to learn BNs. The learning algorithm's goal is to create BNs for the target task.

$$BN^t, \text{from dataset } D^k \left(k = 1...,n\right)$$

Phase 1: To perform a conditional independence test, we used mutual information and conditional mutual information. (CI test), for k∈ [1,n], k≠ t:

$$S_k(A,B \mid C) = S_k^g * S_k^l(A,B \mid C)$$

The expression given above shows the similarity of distribution between target and auxiliary dataset defined by the product of global similarity $S_k^g$ and local similarity $S_k^l(A,B \mid C)$.

Phase 2: This will find the best graph in the search space. Here we have extended the score function to change the learning scenario.

$$Sc_t(G) = \sum_{k=1}^{n,k \neq t} \alpha_k Sc_k(G \mid D^k)$$

**Parameter Learning**: To finish the parameter learning, instance transfer is used. The data set from the m most proximal auxiliary land square (the land square that corresponds to m highest) was transferred to learn the parameter with target structure Gt. Traditional parameter learning algorithms, such as maximum likelihood estimation, can be used with the data.

$$\theta^* = \arg\sup L(\theta \mid D)$$
$$\theta_{i,j,k}^* = \frac{N_{i,j,k}}{N_{i,j}} \; if \; N_{i,j} > 0$$
$$= \frac{1}{r_i}, else$$

Here is the weight factor of land square, which depends on the following aspects:

➢ Number k dataset size

➢ Distribution similarity between target and auxiliary dataset

➢ Geographical position of the land square.

**Conclusion:** The main aim of the experiments was to show how BNs learn from the rare data of the target class, which can be improved with similar classes of data. To compare the learning outcome, the selected subgroups were identified as targets (shortage of training data), the learning transfer method and the traditional way in which it was created, the 4-edge lost structure found in the traditional method. According to domain experts, missing edges are necessary and results are more accurate. The algorithm incorporates structural learning and parameter learning, taking into account the similarities between the learning functions and the geographical location of the land class. By using data from related databases, micro-network performance has been improved. The stability test results show a significant improvement in terms of structure and parameters when transferring information between single soil measuring functions.

**Another example is data processing techniques** used for machine learning and to improve soil classification and accuracy through machine learning algorithms, such as crop selection and crop assessment, disease forecasting, weather forecasting, discounted pricing and irrigation systems. The proposed system uses the J48 seed separation algorithm and is based on predicting soil toxicity levels because accuracy is better than other classification algorithms [21].

The main objectives of the proposed system are:

➤ Monitor crop life based on soil fertility and recommend needed fertilizers

➤ Predicting soil toxicity so that farmers can take action

➤ Raising awareness of the need for irrigation

➤ Crop recommendation.

**Proposed System:** Set of data: For the recommendation system, as well as the final output, two data sets are created. 1. Crop Data Collection: The crop name, temperature, humidity, pH, and nutrients are all included in this crop data set (Na, Mg, N, P, Cl, Ca). 2. Fertilizer Data Collection: The fertilizer name, crop name and date are all included in this data set.

**Hardware Required:** Basic architecture of proposed system is shown in Figure 8.5. The hardware required are: Raspberry Pi 3, Grove SHT 10, Nanotube Sensor, Solar Panel, GSM SIM 300 module and so on.

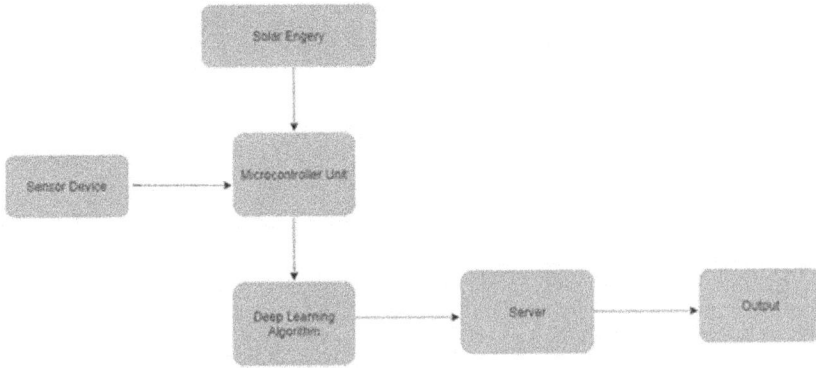

FIGURE 8.5    Architecture of proposed system.

## 8.8.1 Methodology

Decision Tree: A specific instance of the graph space is the Decision Tree. It can be used to model problems when a series of choices leads to a solution. A proposed answer to the challenge of getting from the tree's root to its leaves. The trees are built from the top down. It's utilized to increase the algorithm's forecasting and classification accuracy, as well as a wide range of other applications requiring stable data sizes or distribution. ID3, C4.5, C5.0, and J48 are some of the decision tree algorithms available. J48 algorithm includes the following steps. If instances belong to the same class, the tree represents it as a leaf node by labelling it with the same class. For each attribute given by test on attribute, the potential information is determined. The test on attribute yields the information benefit. On the basis of the current selection criterion, the best spilt is identified and chosen for branching. Figure 8.6 shows the flow diagram of the proposed system.

The proposed approach helps farmers learn about farmland without visiting laboratories and tells them about the fertilizers they should use to increase yields. With the help of key methods such as data entry, it has become much easier to classify the soil based on fertility parameters. Sensor accuracy is very important. The Decision Tree algorithm has a higher accuracy than other classification algorithms. With pollution levels rising rapidly within the industry today, the proposed system will inform farmers about the level of toxins in their soil. If there is poison in the farmer or the consumer, they can take appropriate measures to strengthen their land. This project helps Indian farmers to learn about soil on their land and increase crop yields.

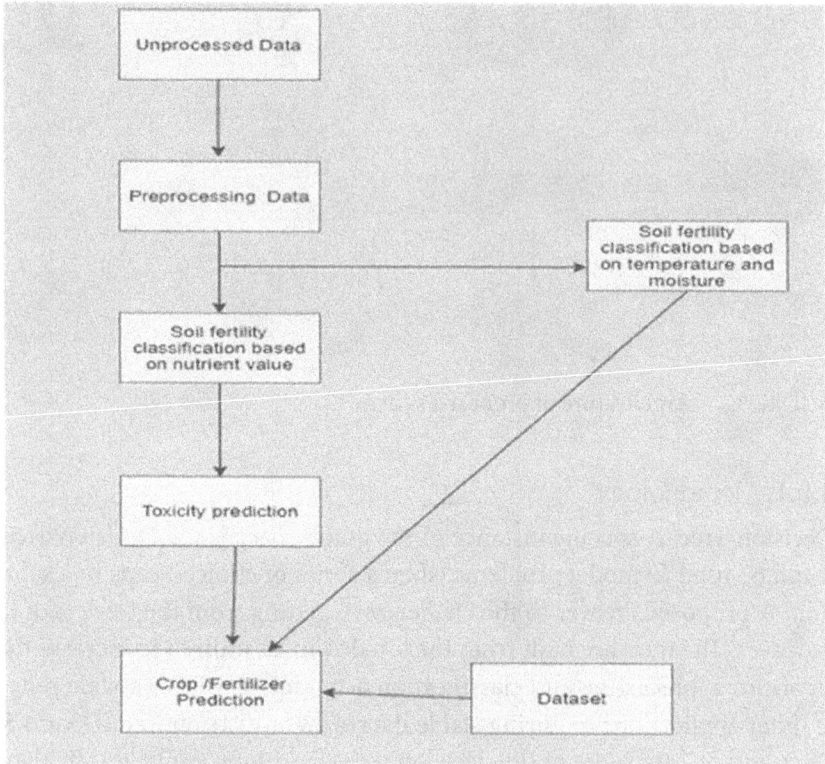

FIGURE 8.6   Crop/fertilizer recommendation system flow diagram.

## 8.9 WEATHER FORECASTING USING DEEP LEARNING AND MACHINE LEARNING FOR AGRICULTURAL MODERNIZATION

Weather prediction is the process of predicting future weather conditions based on previously acquired data. Predicting weather conditions is difficult because it is dependent on a variety of factors. Weather conditions will change every few hours and might be rather dramatic at times. Knowing the weather conditions ahead of time can help us save money in a variety of ways. Its applications range from assisting a student in carrying an umbrella when it was predicted that it would rain in the evening to assisting government organizations in evacuating residents of a location when it was predicted that there would be significant rain in the area. Traditional weather forecasting systems, which rely on satellite pictures and weather stations, are costly due to the employment of costly and complex methods.

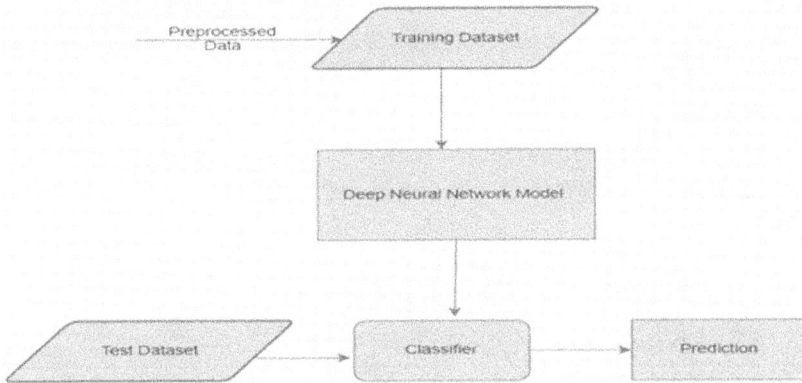

FIGURE 8.7    Weather forecasting system architecture using deep learning.

Recent studies have found machine learning-based weather prediction is less expensive, less time-consuming, convenient, real-time and accurate. Using a machine learning technique to predict weather conditions necessitates the utilization of a large amount of historical weather data. The accuracy of the forecasts is only as good as the data used to train the models. As a result, any machine learning model must be trained with extremely accurate data. Data acquired from a variety of sources is not necessarily reliable. As a result, the data must be pre-processed. Pre-processing data includes deleting columns that aren't essential to the model's predictions, removing zero values, combining related columns, and a variety of other operations. Some prior machine learning-based weather prediction experiments have yielded remarkably accurate results. Several machine learning and neural network models are used in these works. These models have had a great deal of success in predicting values that are extremely near to those predicted by physical and mathematical models. Knowing the weather forecast ahead of time allows us to plan our day more efficiently and respond appropriately. Weather forecasting is challenging due to its irregular and fluctuating nature. In using machine learning to accurately anticipate weather conditions such as temperature, humidity and precipitation, we generally use the following steps [26]. Figure 8.7 shows the architecture of a weather forecasting system.

Data Collection: This contains all of the weather-related data.
Data Pre-Processing and Feature Extraction: All of the data set's properties are analyzed once, and the most important features are extracted to be used later.

Deriving Another Data Set: Another data set was created by averaging the temperature, humidity, pressure and rain for a certain date from the initial data set.

Training and Testing the Machine Learning Model: The random forest classification model is trained with 75% of the resulting data set, with the remaining 25% serving as the model's test set.

## 8.10  DETECTION OF LEAF DISEASE USING MACHINE LEARNING AND DEEP LEARNING

Agriculturists in provincial areas may believe it is difficult to distinguish between the diseases that can be present in their harvests. It is not reasonable for them to visit the agricultural office to determine the nature of the illness. Our main goal is to identify the disease that has been introduced into a plant by observing its shape using image processing and machine learning.

Pests and diseases cause crops to be destroyed or parts of plants to be destroyed, resulting in lower food production and in food insecurity. In addition, knowledge of pest management or control, as well as diseases, is lacking in a number of developing countries. Toxic infections, poor disease control and extreme climate change are just a few of the contributing factors indwindling food output. A variety of current technologies have arisen to reduce post harvest processing, improve agricultural sustainability, and increase output. Various laboratory-based methodologies for illness identification have been used, including polymerase chain reaction, gas chromatography, mass spectrometry, thermography and hyper spectral techniques. These methods, however, are inefficient and time-consuming [15, 28].

In recent years, a server-based and mobile-based approach to disease identification has been used. Several features of these technologies, including a high-resolution camera, high-performance processing, and a large number of built-in accessories, contribute to autonomous disease recognition.

Generally, we adopt three steps for finding the leaf disease using machine learning.

**Dataset:** For this we have to analyses thousands of images of plant leaves and create a dataset of different diseases and healthy leaves.

**Performance Measurement:** To obtain a feel of how our approaches will perform on new data that hasn't been seen before, as well as to keep track of whether or not any of our approaches are overfitting, we evaluate the applicability of machine learning method for the classification problem that is the detection of leaf dieses.

Plants are a significant wellspring of nourishment for the total populace. Plant diseaseadds to the creation of misfortune, which can be handled with persistent observation. Manual plant illness checking is both difficult and inclined towards mistakes. Early identification of plant illnesses utilizing PC vision and artificial intelligence can assist with decreasing the unfavorable impacts of sicknesses, and furthermore defeat the weaknesses of persistent human checking. In this work, we propose the utilization of a profound learning engineering dependent on a new convolutional neural organization called Efficient Net on 18,161 plain and sectioned tomato leaf pictures to order tomato infections [10,27]. The presentation of two division models, U-net and Modified U-net, for the division of leaves is accounted for. The similar performance of the models for binary classification (healthy and unhealthy leaves), six-classclassification (healthy and various groups of diseased leaves), and ten-classclassification (healthy and various kinds of unfortunate leaves) are additionally announced. The modified U-net division model showed precision, IoU, and Dice score of 98.66%, 98.5% and 98.73%, individually, for the division of leaf pictures. EfficientNet-B7 showed prevalent execution for the paired classifification and six-class classification utilizing sectioned pictures with a precision of 99.95% and 99.12%, separately. Finally, Efficient Net-B4 achieved an accuracy of 99.89% for ten-classclassification using segmented images. It may very well be presumed that every one of the designs performed better in grouping the infections when prepared with more profound organizations on portioned pictures. The presentation of every one of the exploratory investigations detailed in this work beats the current writing.

## REFERENCES

1.  M. K. Dharani et al. (2021). Review on crop prediction using deep learning techniques. *Journal of Physics: Conference Series* 1767, 012026 IOP Publishing doi:10.1088/1742-6596/1767/1/012026.

2.  Das, S. D., Deb, N., Biswal, G. R., &Das, S. (2019). *High Voltage Aspects of Smart Agriculture through GIS towards Smarter IoT.* Paper presented at the 2019 International Conference on Automation, Computational and Technology Management (ICACTM).

3. Patel, G.S., Rai, A., Narayan Das, N., &Singh, R.P. (Eds.). (2021). Smart Agriculture: Emerging Pedagogies of Deep Learning, Machine Learning and Internet of Things. CRC Press. https://doi.org/10.1201/b22627.

4. Ale, L., Sheta, A., Li, L., Wang, Y., &Zhang, N. (2019). *Deep Learning Based Plant Disease Detection for Smart Agriculture.* Paper presented at the 2019 IEEE Globecom Workshops (GC Wkshps).

5. Andreas Kamilaris and Prenafeta-Boldú, Francesc X. (2018). Deep learning in agriculture: A survey. *Computers and Electronics in Agriculture*, 147, 70–90, ISSN 0168-1699, https://doi.org/10.1016/j.compag.2018.02.016.

6. Barth, R., Ijsselmuiden, J. M. M., Hemming, J. & Van Henten E. J. (2017). Optimising realism of synthetic agricultural images using cycle generative adversarial networks. *Proceedings of the IEEE IROS workshop on Agricultural Robotics*, Kounalakis, Tsampikos, van Evert, Frits, Ball, David Michael, Kootstra, Gert and Nalpantidis, Lazaros, Wageningen: Wageningen University & Research, pp. 18–22. http://library.wur.nl/WebQuery/wurp ubs/ 533105.

7. Hassina, Ait Issad, Rachida Aoudjit, Belkadi, Malika, Lalam, Mustapha and Daoui, Mehammed (2019) Deep learning-based crops and weeds classification in smart agriculture. *International Journal of Psychosocial Rehabilitation* 23(2), 428–438.

8. Smith, M. L., Hansen, M. F, Smith, L. N., Salter, M., Baxter, E. M., Farish, M. & Grieve, B. (2018). Towards on-farm pig face recognition using convolutional neural networks. *Computers in Industry* 98, 145–152.

9. Jia. Hai-Yanget al. (2010).Soil fertility grading with Bayesian Network transfer learning, *Proceedings of the Ninth International Conference on Machine Learning and Cybernetics*, Qingdao.

10. Chowdhury, M.E.H, Rahman, T., Khandakar, A., Ayari, M.A., Khan, A.U., Khan, M.S., Al-Emadi, N., Reaz, M.B.I., Islam, M.T. and Ali, S.H.M. (2021). Automatic and reliable leaf disease detection using deep learning techniques. *AgriEngineering* 3, 294–312. https://doi.org/10.3390/agrieng ineering3020020.

11. Demmers, T.G., Cao, Y., Parsons, D.J., Gauss, S. and Wathes, C.M. (2012). Simultaneous monitoring and control of pig growth and ammonia emissions. *IX International Livestock Environment Symposium (ILES IX).*American Society of Agricultural and Biological Engineers, Valencia, Spain.

12. Jia, Y., Shelhamer, E., Donahue, J., Karayev, S., Long, J., Girshick and R. Darrell, T. (2014). Caffe: Convolutional architecture for fast feature embed-ding. *Proceedings of the 22nd International Conference on Multimedia*, pp. 675–678). Orlando, FL: ACM.

13. Rai, Amrita, Sharma, Deepti, Rai, Shubhyansh, Singh, Amandeep and KantSingh, Krishna (2021). "IoT-aided robotics development

and applications with A'" in Kant Singh, Krishna, Nayyar, Anand, TandarSudeep and Abouhawwash Mohamed. *Emergence of Cyber Physical System and IoT in Smart Automation and Robotics*. Cham: Springer.

14. Bu, F., & Wang, X. (2019). A smart agriculture IoT system based on deep reinforcement learning. *Future Generation Computer Systems* 99, 500–507.

15. G. Rutu et al. (2018). Plant disease detection using CNNs and GANs as an augmentative approach, *IEEE International Conference on Innovative Research and Development (ICIRD)*.

16. Khaki Saeed and Wang Lizhi (2020). Archontoulis Sotirios: A CNN-RNN Framework for Crop Yield Prediction Frontiers in Plant Science, 10, p. 1750. www.frontiersin.org/article/10.3389/fpls.2019.01750.

17. Keerthan Kumar, T. G., Shubha C. and Sushma, S. A. (2019). Random forest algorithm for soil fertility prediction and grading using machine learning, *International Journal of Innovative Technology and Exploring Engineering (IJITEE)* ISSN: 2278–3075, 9(1).

18. Rusia, K. Rai, S. Rai A. and Kumar Karatangi, S. V. (2021). "Artificial intelligence and robotics: Impact &open issues of automation in workplace", *2021 International Conference on Advance Computing and Innovative Technologies in Engineering (ICACITE)*, pp. 54–59, doi: 10.1109/ICACITE51222.2021.9404749.

19. Li, G., Huang, Y., Chen, Z., Chesser, G.D., Jr., Purswell, J.L., Linhoss, J. and Zhao, Y. (2021). Practices and applications of convolutional neural network-based computer vision systems in animal farming: A review. *Sensors*, 21, 1492. https://doi.org/10.3390/s21041492.

20. Panchamurthi, S. (2019). Soil analysis and prediction of suitable crop for agriculture using machine learning. *International Journal for Research in Applied Science and Engineering Technology* 7, 2328–2335. 10.22214/ijraset.2019.3427.

21. Mayuri, P. and Geetha, C. (2018). Soil toxicity prediction and recommendation system using data mining in precision agriculture, *3rd International Conference for Convergence in Technology (I2CT)*.

22. Pan, S.J. and Yang, Q. (2010). A survey on transfer learning. *IEEE Trans. Knowl. Data Eng.* 22 (10), 1345–1359.

23. Barth, R., Ijsselmuiden, J., Hemming, J. and Van Henten, E.J. (2019). Synthetic bootstrapping of convolutional neural networks for semantic plant part segmentation, *Computers and Electronics in Agriculture* 161, 291–304, ISSN 0168-1699, https://doi.org/10.1016/j.com pag.2017.11.040.

24. Song, X., Zhang, G., Liu, F., Li, D., Zhao, Y. and Yang, J. (2016). Modeling spatio-temporal distribution of soil moisture by deep learning-based cellular automata model. *Journal of Arid Land* 8(5), 734–748.

25. Sehgal, G., Gupta, B., Paneri, K., Singh, K., Sharma, G. and Shroff, G. (2017). Crop planning using stochastic visual optimization. arXiv preprint arXiv: 1710.09077.

26. Nitin, S. et al. (2019), *Weather Forecasting Using Machine Learning Algorithm*, https://ieeexplore.ieee.org/xpl/conhome/8930208/proceeding

27. Venkat Narayana Rao, T. and  Manasa, S. (2019). Artificial neural networks for soil quality and crop yield prediction using machine learning. *International Journal on Future Revolutionin Computer Science & Communication Engineering* 5(1), ISSN 2454-4248.

28. Hardikkumar, S. et al. (2020). Plant leaf disease detection and classification using conventional machine learning and deep learning, *International Journal on Emerging Technologies* 11(3), 1094–1102.

# CHAPTER 9

# Applications of Deep Learning in Aerial Robotics

Prabhu Jyot Singh

*Central Queensland University, Sydney, p.singh@cqu.edu.au)*

## 9.1 INTRODUCTION

Nowadays, the aerial robotics industry is designing and developing various powerful aerial robots. These robots are used in different application scenarios such as agriculture, disaster management, search and rescue operations and surveillance. The robots can perform better operations if they can work autonomously without the need for human interpretation. For this purpose, these aerial robots need to be trained through a deep learning algorithm that relies on image and sensor data. In this chapter, the various deep learning applications of aerial robotics are discussed based on image and sensor data. The deep learning applications are also categorized based on the different functions performed by these aerial robots. The chapter starts with a discussion of the basic principle of aerial robotics to understand how aerial robotics performs its functions, and then goes into detail about the deep learning application of aerial robotics.

The remainder of the chapter is as follows. Section 9.2 gives an overview of aerial robotics, covering the basic principles and functions. Section 9.3 categorizes different aerial robots based on flight time, payload and communication range. This section summarizes the different types of aerial

DOI: 10.1201/9781003217237-9

robots that are available in the market. Section 9.4 explores the different applications of aerial robots. The farming applications of aerial robots are presented in Section 9.4.1. Section 9.4.2 covers the logistic applications of aerial robots. The surveillance applications of aerial robots are mentioned in Section 9.4.3, and natural disaster applications are covered in Section 9.4.4. The role of deep learning in aerial robots is discussed in Section 9.5. The architecture of aerial robots is presented in Section 9.6. This architecture is used to understand the different components where the deep learning algorithms can be applied to make aerial robots more powerful for performing autonomous operations. In Section 9.7, the deep learning applications of aerial robots are discussed. These deep learning applications are categorized based on the four components of aerial robots, namely: Featured Extraction, Planning, Motion control and Situation Awareness. In this chapter, three learning (supervised, unsupervised and reinforcement) methods with the different algorithms are covered, which are based on image and sensor data. These deep learning applications are summarized in Section 9.7.5. Finally, the chapter is concluded in Section 9.8.

## 9.2 OVERVIEW OF AERIAL ROBOTICS

New technology is growing all around us and becoming part of our day-to-day life. Aerial robotics is one of these technologies. As the name suggests, an aerial robotis a robotics instrument that can fly in the air. Aerial robots are also known as unmanned aerial vehicles (UAVs) or drones. There are lots of aerial robots available on the market, which can be classified according to their size, weight and payload capacity. Based on their size, they are known as mini-UAV, micro-UAV, small size UAV and large size UAVs. In this chapter, the terms drones, aerial robots and UAVs are used interchangeably. Before going further into a deep discussion about the application of aerial robotics, it is necessary to understand their functionality. The latest model of aerial robotics contains one mechanical object with rotors, a control circuit and a battery. A controller is required to control these aerial robots. These aerial robots are controlled by a smartphone mobile application and are known as wi-fi operated aerial robots. Aerial robots are being used in various commercial applications related to our day-to-day life. For example, some of these applications are farming, parcel delivery, natural disaster, surveillance and monitoring.

## 9.3  CLASSIFICATION OF AERIAL ROBOTICS

Aerial robots can be classified in different ways. As per [1], there are two types of aerial robots: heavier-than-air and lighter-than-air. The heavier-than-air can be subdivided into two categories based on the wing type and rotor type. Fixed-wing, flying-wing and flapping-wing are categorized as under wing type, and helicopter, quadcopter, hexacopter and octocopter are considered as rotor type heavier-than-air aerial robots. Blimp and balloon are two types of lighter-than-air aerial robots. These aerial robots can also be differentiated based on the flight time, payload, degree of autonomy and degree of sociability. In the other classification of air robotics, Watts et al. [2] categorise them based on the flight capacity and capabilities. In their research, they discussed different types of air robotics based on civil and military applications. These aerial robots are Micro or Miniature Air Vehicle (MAV), Nano Air Vehicle (NAV), Vertical Take-Off & Landing (VTOL), Low Altitude Short Endurance (LASE), Medium Altitude Long Endurance (MALE) and High Altitude Long Endurance (HALE). In research conducted by Gupta et al. [3], they classified aerial robots as MALE, HALE, Tactical UAV (TUAV), Mini UAV (MUAV), MAV and NAV. Brooke-Holland [4] also classified aerial robots into three different classes. Class I consist of Nano, Micro, Mini and small aerial robots; tactical aerial robots come in Class II; and Class III has MALE, HALE or Strike aerial robots. In other research on the classification of aerial robotics, Arjomandi et al. [5] highlighted the different types of aerial robots. They categorised air robots based on the weight into five categories: Superheavy (with weight more than 2000 kg), Heavy (with weight more than 200 kg and less than or equal to 2000 kg), Medium (with weight greater than 50 kg and less than or equal to 200 kg), Light (with weight between 5 and 50 kg) and Micro (with weight less than 5 kg). Cavoukian [6] classified aerial robots as micro, mini and tactical aerial robots. Weibel and Hansman also classified air robots based on the weight type. They categorized them as micro, mini, tactical, medium and high altitude and heavyweight aerial robots [7]. Aerial robots were also classified by different countries' Civil Aviation Safety Authority (CASA). As per the UK Civil Aviation Authority: small aerial robots that weigh less than or are equal to 20 kg, light aerial robots with the weight of between 20 and 150 kg and aerial robots with weight over 150 kg [8,9]. The Australian CASA also categorises aerial robots as micro, small and large. The micro aerial robots have a weight less than 0.1 kg, small aerial robot weight between 0.1 and 150 kg, and over 150 kg air robots are considered as

TABLE 9.1   Classification of aerial robots

| S. no. | Aerial Robot | Uses | Weight |
|---|---|---|---|
| 1 | Super Heavy | Military use | Weight usually, more than 2000 kg |
| 2 | MALE | Military use | More than 600 kg |
| 3 | HALE | Military use | More than 600 kg |
| 4 | Tactical | Aerial photography and cinematography, to intelligence, mapping, | Weight lies between 150 and 600 kg |
| 5 | Small | Intelligence gathering, scientific studies and meteorological research | Weight between 20 kg and 150 kg |
| 6 | Mini | Intelligence gathering, environmental monitoring | Weight between 2 kg and 20 kg |
| 7 | Micro | Environmental monitoring, surveillance or disaster management | Weight between 200g and 2 kg |
| 8 | Nano | Reconnaissance, surveillance and target acquisition | Weight less than 200 g |

large [10]. The aerial robots were also classified based on their mission such as attack aerial robot application, attack aerial robot expendable, strategic, tactical and miniature aerial robot [11]. Zakora and Molodchik [12] also classified aerial robots as combat, heavy large endurance, heavy medium-range, medium-heavy, average, lightweight medium, lightweight small range, micro and mini close range.

Different authors have categorized aerial robots based on different parameters, such as flight time, payload capacity, weight and range of communication, but most of them have categorized aerial robots as nano, mini, small tactical and big size. A summary of different types of aerial robot is given in Table 9.1.

## 9.4 APPLICATION OF AERIAL ROBOTICS

Initially, aerial robots were used for military and government organizations. But a huge development in aerial robots has moved them into the commercial market. Today, the use of aerial robots is very common in various commercial applications, such as agriculture, logistics and transportation, surveillance and in natural disasters. These are a few of the most common aerial robots application for commercial purposes. In the next subsection, these applications of aerial robots are discussed in turn.

### 9.4.1 Use of Aerial Robotics for Farming

Aerial robots are being used in various farming applications. Farmers can monitor their farms with the help of an aerial robot. In the agriculture sector, aerial robots can be used to pesticide the crops and to find out the growing characteristics of the crops in different weather conditions. There are several applications of aerial robots in the field of agriculture and farming. Saari et al. reviewed 100 applications of aerial robots in [13] for precision agriculture. In their research work, the most common applications of aerial robots for precision agriculture was discussed: as crop spraying, irrigation management, weed mapping, vegetation growth and health monitoring, disease detection and yield estimation. In [14] the author discusses 20 different aerial robotic applications of crop monitoring or spraying processes for precision agriculture.

Air robots will be more common in the near future, and will be used more by individuals or farm owners for precision agriculture applications.

### 9.4.2 Use of Aerial Robotics for Logistics

Another application of aerial robots is in the field of logistic management. Aerial robots were tested to deliver parcels. There was not a huge success rate, due to constraints like payload, route direction and other difficulties, but it will be a major use of aerial robots in the future for commercial purposes. In [15], the author discusses technology challenges for parcel delivery by aerial robots that depends on their efficiency, coordination and artificial intelligence. Many research groups are working on this issue, and it will be a more effective way to deliver a parcel through an aerial robot in the near future. The CASA Australia approves Wing Aviation PTYLtd to operate aerial robots' delivery in North Canberra (ACT) and Logan (QLD) [16]. Most countries are testing aerial robots for delivery purposes and working on the challenges and issues but as per the current research and development in aerial robots, the day will soon come when these aerial robots will be the common medium of delivery.

### 9.4.3 Use of Aerial Robotic for Surveillance

It is very common to use aerial robots for surveillance. Puri researched the use of aerial robots for traffic surveillance. He presented a survey report in [17] and discussed the application of aerial robots for traffic surveillance. In [18] Finn and Wright discuss the aerial robot's surveillance system. They discuss the use and impact of aerial robots for surveillance in civil applications.

### 9.4.4 Use of Aerial Robotics for Natural Disaster

Aerial robotics is helping in natural disasters as well. There are lots of applications where aerial robotics were tested for natural disaster. They can be used for earthquakes, floods and bush fires. In research [19] Erdelj and Natalizio discuss the aerial robot's application for disaster management. They measure the performance of live video streaming applications for disaster management. In [20] Estrada and Ndoma discuss the use of aerial robots for a natural disaster. As a recent advanced development of aerial robots, they are very useful in the event of a natural disaster and will save lots of lives in future.

## 9.5 ROLE OF DEEP LEARNING IN AERIAL ROBOTICS

Artificial intelligence used with aerial robotics makes them a more powerful object for use in the commercial market. If an aerial robot can take the decision for its route without any collision then it will solve lots of the problems. Deep learning algorithms were used to train the aerial robot for autonomous route direction by providing them with sensor and image data. Aerial robots can use supervised, unsupervised or reinforcement learning algorithms for autonomous operations. In future, when lots of UAVs are flying in the sky for various commercial applications, they will need to operate autonomously based on sensor and image data to avoid a collision. Deep learning algorithms will help to work aerial robots as autonomous objects and they can take a decision for avoiding collision with other aerial robots in the swarm of robots. Deep learning in aerial robotics makes aerial robots a more powerful tool in various commercial applications. Air robotics automation solves various problems and has been used in various applications. In past research, air robotics automation was used in different civil applications such as inspection of power lines [21]. In other research, carried out by Olivares et al. [22], they discussed air robotics automation application for wildlife conservation. Similarly, air robotics automation application for building inspection and precision agriculture is also covered in past literature such as [23] and [24].

## 9.6 AIR ROBOTS' ARCHITECTURE AND COMPONENTS

It is an interesting question, how an aerial robot can be trained and how a deep learning algorithm can be implemented to make it a more powerful tool with automation. In this process, the first task is to understand the architecture of an aerial robot, which will help to explore in detail on which

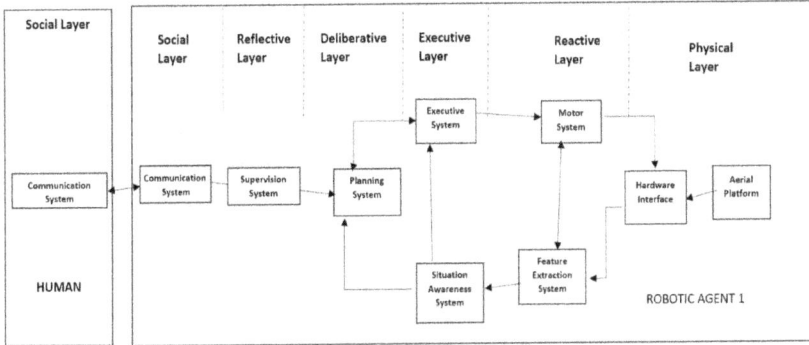

FIGURE 9.1    Aerostack multilayer architecture.

Source: Sanchez-Lopez et al. [25]

components of aerial robots deep learning algorithms can be applied and utilized their functionality in various aerial robotics application domains. Sanchez-Lopez et al. [25] presented a multilayer aerostack architecture that has several components. This architecture is shown in Figure 9.1.

First, we briefly discuss the different components of this architecture. It will also help to understand how aerial robots navigate, communicate and perform other operations. With an understanding of these components, the deep learning algorithm for aerial robots can be analysed.

As per Figure 9.1, there are mainly eight components of aerial robot architecture: communication system, supervision system, planning system, executive system, situational awareness system, motor system, feature extraction system and hardware interface. These componentsare discussed one by one in the section below to get a clear idea of aerial robots' functions.

> **Communication System**: This includes the components that are used for maintaining communication between aerial robots with human operators or with other robots.
>
> **Supervision System**: This component belongs to the supervision system related to intelligence. These are self-awareness components that can command other systems. A deep learning algorithm can be used to evaluate how this component works towards achieving the goal while there are lots of obstacles present and how to recover from those problems.
>
> **Planning System**: One of the most important components of aerial robots is their planning system. This component is related to the path planning from source to destination as well as the mission planning.

**Executive System**: This system generates a behaviour sequence that is based on the input of high-level symbolic action.

**Situational Awareness System**: This system also includes important components for aerial robots. An aerial robot has lots of sensors installed with its hardware and continuously receives data from its sensor. The component of this system generates state variables information related to the robot and its environment from the sensor information.

**Motor System**: This system is responsible for the control of position, seep and orientation of the aerial robot. Motion controllers are the main component of the motor system that receives commands to control the robot. The actuators receive these commands after these commands are converted into low level commands.

**Feature Extraction System**: The extraction of meaningful features or representations from sensor data is referred to as feature extraction. Because most deep learning algorithms are tasked with learning data representations, feature extraction techniques are inextricably linked to them.

**Hardware Interface**: The component for the hardware system includes the actuators and sensors.

The architecture of air robots gives a brief idea about its components that can be used to discuss in detail the application of deep learning in aerial robotics. Deep learning applications of aerial robotics are mostly based on planning, situation awareness and motion control. These applications are covered in the next section.

## 9.7 APPLICATION OF DEEP LEARNING IN AERIAL ROBOTICS

### 9.7.1 The Deep Learning Application for Feature Extraction in Aerial Robotics

As discussed in the previous section, the feature extraction system is directly related to the sensor of the aerial robot. Aerial robots have lots of sensors where they can get the data to the onboard robot system. There is a wide range of applications of deep learning for feature extraction, which is related to the image sensor of aerial robots. These image sensors used different techniques such as monocular RGB camera, infrared and RGB-D sensors for the feature extraction system. But the deep learning for feature extraction is mostly based on convolutional neural networks (CNNs) [26]. In aerial robotics, feature extraction techniques based on CNNmodels have primarily been used in the recognition of objects.In [27] Girshick et al. presented a simple and scalable object detection algorithm for aerial robots, and the Fast Region-based CNN method for object detection was discussed in [28]. In another research [29] based on the CNN model, Ren et al. presented a Region Proposal Network (RPN). This RPN was used for full picture convolutional highlights with the detection network.A CNN model was used with RPN for the objectness scores as well as for the prediction of objects. Lee et al. [30] also used the R-CNNs' algorithm in real-time Cloud-based object detection for aerial robots. Another approach of object detection was presented by Redmon et al. [31]. They proposed a You Only Look Once (YOLO) system for object detection, based on the single regression problem. This YOLO system allows prediction of an object that was present and their location after only looking at them once. In their continuous research, Redmon and Farhadi proposed the new version of the YOLO system known as YOLO9000. It was a better, stronger system for real-time object detection, which was able to detect over 9000 object categories [32]. Liu et al. [33] presented a method for object detection in images known as Single Shot MultiBox Detector by using a single deep neural network. The usage of multi-scale convolutional bounding box outputs that bind with the various features maps atthe top of the network was a crucial aspect of this model. In most of these object recognition techniques, object classification and bounding box regression were combined with the CNN model. Another approach can be used for object detection by aerial robots with the help of unsupervised learning. This approach was used in [34], with less requirement on manually training data. But this approach was more costly and time-consuming.

Scene classification is another area of feature extraction system in aerial robotics, which is also based on the CNN model. Research proposed by Zhou et al. in [35] presented a new scene-centric database called Places, which had more than seven million labelled pictures of scenes. In this method, they compared the density and diversity of image datasets and proved that Places had more diversity. Aerial and remote sensing image classification was done by Penatti et al. in [36], based on CNN. They experimented and evaluated the result with trained CNN for everyday object recognition. In this experiment, they also compared the CNN with other visual descriptors. These experimental results showed the highest accuracy rate for aerial images with CNN. In [37], Hu et al. discussed how to transfer features from pre-trained CNNs for scene classification. They proposed two scenarios for extracting CNN features from different layers to generate picture features in this research. The activation vectors were used in the first scenario to get the final picture features and these vectors were regained from the fully connected layers. The global image feature was retrieved from the dense image feature in the second scenario. This dense image feature was based on the last convolutional layer. In other research for the dynamic scene classification, statistical aggregation (SA-CNN) was used to solve the problem of dynamic video capturing [38]. Scene classification was experimented with using the quadrotor micro aerial vehicle for the visual perception of forest trails in [39]. In this experiment, an image of forest trails was classified into three classes (turn left, go straight and turn right) with the help of ten layered CNN models to provide the instruction to the aerial robot to maintain itself on the trail.

The object detection and scene classification from the aerial robotic images with deep learning techniques have various applications in different domains. The agricultural application domain is one of them. There are lots of applications of aerial robotics for the agricultural domain with and without deep learning techniques. The common application of aerial robots is to capture live images of farmland and send them back to the ground control station. But with deep learning, object detection and scene classification techniques of aerial robots make themselves more powerful in the agriculture sector. The most useful applications of deep learning of aerial robotics in the agricultural field are plant identification, counting, monitoring the state of crops, weed classification, crop classification and weed scouting.

Research presented by Li et al. [40], discusses the use of deep learning algorithms for selective weed control with a real-time aerial robot. In [41], a deep learning algorithm was used on the oil palm trees for plan tree detection and counting with remote sensing images. Research was conducted for fruit counting with aerial robotics by using the deep learning algorithm in [42]. Hung et al. [43] also presented their research for the application of deep learning in aerial robotics for the agricultural sector. They proposed feature learning for the classification of invasive weed species with the help of small aerial robots. A case study was presented in [44] for crop classification from the high quality images of the aerial robot with the help of a deep learning algorithm. In all of this research, the agriculture application of aerial robotics has a powerful impact with the use of deep learning algorithm and it solved various problems and gave more power to these robots to perform their specific tasks.

The main aim of an aerial robot is to capture live images and videos when it flies in the sky and to send them back to the ground control station. Some other aerial robotics applications use deep learning techniques to capture images for monitoring, search and rescue operations. Sawarkar et al. [45] proposed a teleoperating robot for combat, rescue and reconnaissance missions. This robot has an IP camera that broadcast the live videos to the remote cloud server. With the help of the deep learning algorithm, these videos were used to identify and recognize individuals. In other research [46], avalanche search and rescue operations were performed with an aerial robot that has a fitted vision camera by using deep learning. In this approach, pre-trained CNN was used to extract discriminative features from images of the avalanche debris that were captured by the aerial robot. Kim and Chervonenkis [47] presented their work for road traffic monitoring with anaerial robot by using deep learning that was based on automatic situation control. In this research, they provide asolution for vehicle tracking and detection problems for assisting the traffic situation with the deep learning methods. Kim et al. [48] also discussed the development of an aerial robot type jellyfish monitoring system with the help of a deep learning algorithm.

Object recognition with the help of deep learning in aerial robotics requires a Graphical Processing Unit (GPU) that deals with real-time constraints. There are lots of variables that need to be considered in designing and developing an aerial robot, such as flight time, payload, weight and onboard software. Hardware and software both need to be embedded onboard aerial robots to make them successful in performing various tasks. Due to these constraints, there were limited aerial robotic systems having embedded feature extraction algorithm using deep learning was processed by GPU technology on board of an aerial robot. A similar problem to this was discovered in [49] called Automatic Detection, Localization and Classification (ADLC). This ADLC algorithm combines the CPU cores that allow working GPU on the classification task.

Deep learning for featured extraction in aerial robotics was discussed above and all of them are based on the data captured by the image sensor of the aerial robot. As discussed in the architecture of aerial robot in Section 9.6, other sensors also play an important role in the overall operation of the aerial robot. So, the deep learning techniques can be applied to the other data sensors of aerial robots such as laser sensors, radar sensors and acoustic sensors. There has also been some research done in the past with deep learning techniques with these sensors of aerial robots. In [50], Maturana and Scherer discuss the use of a LIDAR sensor that was mounted on a helicopter to detect safe landing areas. In this research, they useda 3D CNN system for efficient and reliable detection of safe landing zones from the LIDAR sensor. In terms of deep learning techniques on the radar sensor of aerial robots, there has not been much research done in the past. Research presented by the author in [51] used a 2.4 GHz Doppler radar sensor to capture the Spectral Correlation Function that helps to detect micro air robots within three predefined classes. They developed a model that used a semi-supervised deep belief network trained with data for this deep learning purpose. The acoustic sensor of aerial robots collects the acoustic data that is used by deep learning techniques in a different application.Jeon et al. [52] presented their work with the deep neural network for drone sound detection in areal-life environment. They presented a method based on the sound event to detect the presence of a commercial hobby drone. In this research, they investigated the effectiveness of drone sound detection with even sound classification methods such as CNN, RNN and Gaussian Mixture Model (GMM), and found that RNN provides the best detection performance. In other research conducted by Morito et al. [53], a Partially Shared Deep Neural Network (PS-DNN) was proposed with the partially annotated data that was able to deal with the problem of sound source identification and separation. In this research, two overlapped subnetworks of PS-DNN were used. Out of these two networks, one was used for the sound identification as classification network and the other was used for sound source separation as a regression network.

In past research, most of the applications relating to feature extraction belong to the agriculture domain, search and rescue, surveillance and the identification of aerial robots. Most of the research has a CNN algorithm with supervised learning for object recognition and scene classification, which can be used in agriculture applications. However, some agriculture applications use unsupervised learning with autoencoder algorithm. The summary of deep learning for feature extraction is given in Figure 9.2.

### 9.7.2 Deep Learning Applications for Planning in Aerial Robotics

In aerial robotics, deep learning algorithms were used for planning in various applications. When dealing with unstructured, dynamic surroundings or

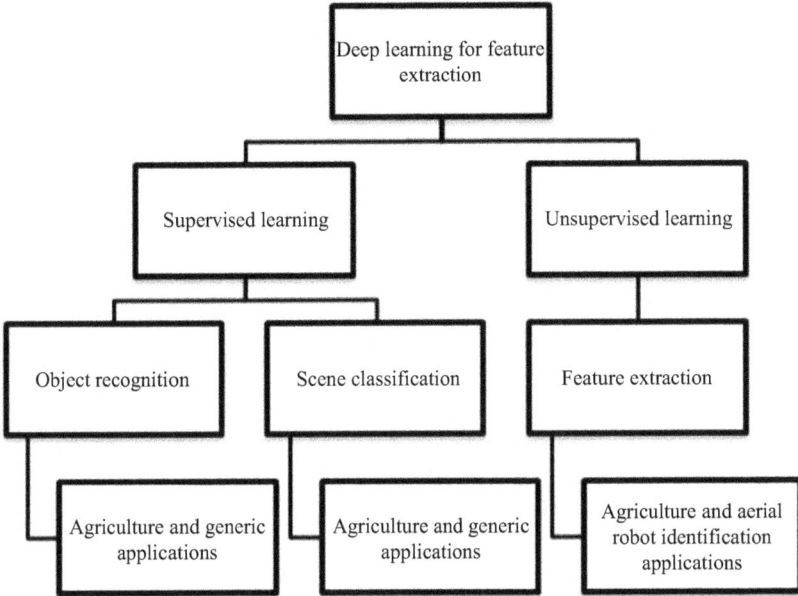

FIGURE 9.2    Deep learning applications for feature extraction.

when the scope and/or tasks of the robot are diverse, planning is necessary. Path motion, navigation and manipulation planning are the common tasks performed by an aerial robot. It will be interesting to understand different tasks that an aerial robot can perform as part of their planning tasks, such as self-localization, mapping and robot state estimation. In [54], Delmerico et al. discussed the active autonomous aerial exploration for ground robot path planning. In this research, they used observations from a flying robot to solve the problem of designing a course for a ground robot through unfamiliar terrain. They proposed active exploration of the environment, in which the flying robot selects places to map in a way that minimized the system's overall response time, which is the time it takes for the air and ground robots to complete their missions together. They used a CNN image classifier for terrain classification and for mapping the landscape and selecting traversable terrain for the ground robot to achieve this goal. In other research conducted by Delmerico et al. [55], they used a system for a vision-based flying robot to lead a ground robot through unfamiliar terrain to a destination location. In this framework of a search-and-rescue air-ground robot team, they devised and verified a system for "on-the-spot" classifier training for terrain mapping. Within one minute after launch in a catastrophe area, an aerial robot can collect training data, train a classifier,

```
┌─────────────────────────┐
│   Deep learning for Path │
│         Planning         │
└─────────────────────────┘
           │
    ┌─────────────────────────┐
    │   Supervised Learning    │
    └─────────────────────────┘
           │
        ┌─────────────────────────┐
        │   Search and Research    │
        │       Applications       │
        └─────────────────────────┘
```

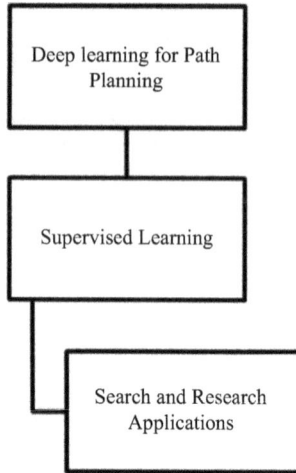

FIGURE 9.3    Deep learning applications for path planning.

and begin terrain mapping in flight. In this research, they used a CNN learning algorithm for search and rescue operations.

Planning is an important task in aerial robotic and the deep learning application of planning makes aerial robots more powerful to perform their operation in different application domains. The summary of the applications of deep learning for planning in arial robots is shown in Figure 9.3.

### 9.7.3 Deep Learning Applications for Motion Control in Aerial Robotics

Motion control is an important feature of aerial robots. An aerial robot gets the motion control commands from the ground control station to fly in the sky along with a specific path. The aerial robot can perform complex manoeuvres based on the control commands and these control commands helped to solve various robotic problems. There are many factors that affect aerial robot manoeuvres, such as wind, rain or damaged propeller. It is important to understand how an aerial robot is able to cope with this situation. These problems can be handled with a deep learning algorithm. In this type of situation, learning with experience comes into the picture and takes an important step to deal with these problems. Deep learning algorithms have the advantage of being able to generalize correctly with certain sets of labelled input data. Deep learning enables the inference of a pattern from raw inputs like pictures and LIDAR sensor data, resulting in appropriate behaviour even in unforeseen scenarios.Recent improvements in aerial robot indoor navigation have resulted in the successful application of CNNs to map pictures to high-level behaviour instructions like turn left, turn right, rotate left and rotate right.

Some research has been done in the past with deep learning algorithms for motion control in aerial robotics. In [56], Kim and Chen discussed the deep neural network for real-time autonomous indoor navigation. In this experiment with Micro Aerial Vehicles (MAVs), they analysed problems for autonomous indoor navigation due to the limited GPS signals as well as the low payload capacity (unable to carry heavy range finder sensors) of MAVs. They used the CNN deep learning model for autonomous indoor navigation of MAVs. In this research experiment, activities were straightforwardly planned from crude pictures. In completely expressed techniques, the learned model was run off-board, normally exploiting a GPU in a laptop device. In other research done by Sadeghi and Levine [57], a learning method for collision-free indoor flight was presented with the real world experiment that was trained on the 3D CAD simulation model. This research shows that an aerial robot can be trained based on the simulation data without considering the real image. They used Q function estimation based on the CNN model that was trained through simulation data for indoor navigation. In [58], navigable zones were projected in the form of up to three bounding boxes from a disparity image.In this experiment, the next waypoint was chosen as the centre of the largest bounding box found and the aerial robot flights were completed using this method. The key disadvantage with this technique is that the disparity pictures must be sent for the computation at the host device. In this process, it took only 1.3 seconds for aerial robot horizontal translation, disparity map development and waypoint selection. Santana et al. presented their research in [59] for swarm-based visual saliency that was related with tracking natural trails. This research suggested a swarm-based visual saliency model capable of embedding prior information of the object's general layout.During the tough task of tracing unstructured paths in natural contexts, the model performed well and quickly. Against a highly challenging and diversified set of conditions, the model has a success rate of 97% at 20 Hz. This method is well suited to visual serving at low speeds, as is common in tight and serpentine paths. Some researchers have focused on congested natural circumstances, such as dense forests or trails, when it comes to aerial robot navigation in unstructured environments. As discussed in [39], a

DNN model with a final softmax layer was used to produce the movement action probabilities (turn left, go straight or turn right) that were based on the mapping images and tested onboard using an ODROID-U3 processor. There is another issue with the aerial robot low-level motion control as it is a big problem to handle continuous and multivariable action space in aerial robots. In [60], Zhang et al. presented their work concerning the low-level motion control for aerial robots. Using model predictive control to create guiding samples for guided policy search, they developed a method for training deep neural network control policies for autonomous aerial vehicles. This Model Predictive Control(MPC)-guided policy search employs a modified MPC method that balances cost minimization and policy matching with the existing neural network policy. Kelchtermans and Tuytelaars [61] altered the well-known Inception v3 model (pre-trained CNN) to allow the final layer to generate six action nodes (three orientations and three transitions). After some retraining, the aerial robot was able to cross a room with a few obstacles strewn around.

Deep learning techniques in motion control for aerial robotics was used in various application domains, such as autonomous navigation of robots. This navigation can be indoor or outdoor with a different type of learning algorithm. For outdoor navigation, supervised learning was used with the CNN model as compared to indoor navigation, where both supervised and reinforcement learning wereused with CNN and GPS algorithms in past research. The summary of deep learning applications for motion control in aerial robots is presented in Figure 9.4.

### 9.7.4 Deep Learning Applications for Situation Awareness in Aerial Robotics

Situation awareness is also an important factor for aerial robots. Deep learning algorithms have been used with situation awareness in aerial robotics in past research. The application of deep learning can be used to localize aerial robots as well as image registration tasks. In [62], research was presented about deep learning for ground-to-aerial glocalization. The author used a reference aerial database and matched the ground level query images with it. In this research, inconsistency between appearance and baseline for the aerial photographs and ground level was overcome with the help of deep learning algorithms.The author used pair based network with unmatched and matched cross view pictures and developed deep representations from these data. The necessity to predict scale, direction

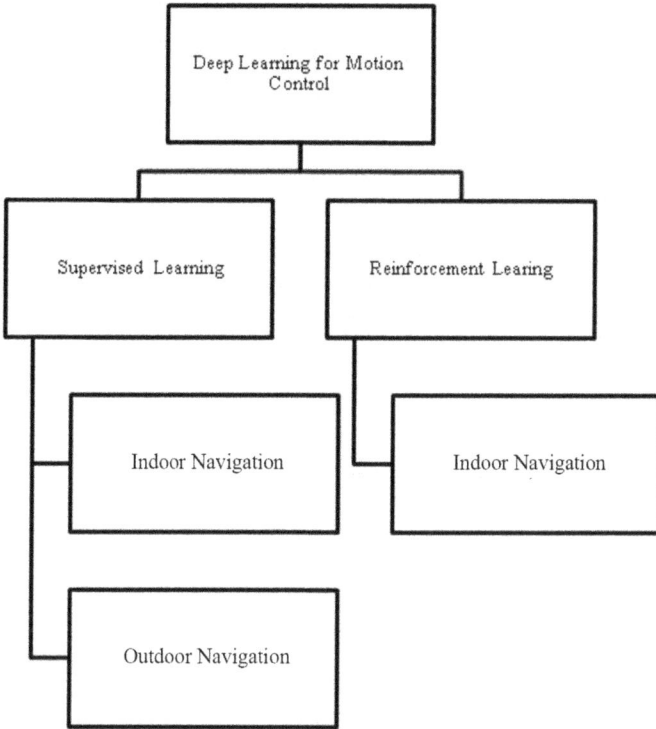

FIGURE 9.4    Deep learning applications for motion control.

and dominating depth at test time for ground-level queries were the key constraints in the ground-to-aerial glocalization research. Taisho et al. presented their work in [63] about the mining visual experience for fast cross-view UAV localization. In this approach, a library of raw image data for the nearest neighbour was used to compare with the input query image for feature extraction. This feature extraction for matching verification was handled by pre-trained CNNs. The retrieval time for the query was not disclosed due to the reduced computational complexity in this approach. Given an image collected onboard and a global motion plan, a CNN was presented in [64] to create movement action plans for an aerial robot. The CNN was used to figure out how to map visuals to position-dependent actions. This was the same procedure for picture registration and control action generation with the global motion plan, but this time CNN was encoded effectively to learn the behaviour and make it superior for image registration techniques. However, these experiments were not conducted with the real aerial robots as well as there was no information

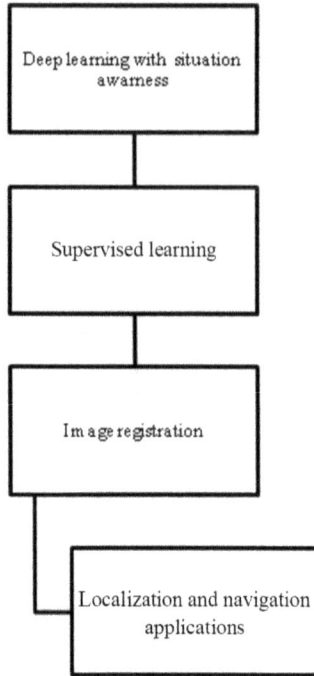

FIGURE 9.5    Deep learning applications for situation awareness.

about the execution time, which could make it more difficult to deploy these techniques in the real aerial robot application.

Most of the applications for deep learning with situation awareness are related with the localization of aerial robots. These applications used a supervised learning model with a CNN algorithm to perform the image registration task. The summary of applications related to deep learning with situation awareness is presented in Figure 9.5.

## 9.7.5 Summary of Deep Learning Applications in Aerial Robots

The various deep learning applications of aerials robotics, which can be used in different fields of application, have been discussed in past research. It was observed that the deep learning application of aerial robotics depends on the learning method for the feature extraction, planning, motion control and situation awareness tasks. Most of the applications used three types of learning algorithms known as supervised, unsupervised and reinforcement learning. These learning algorithms used different sensor technology of aerial robots to perform the specific task based on the trained data.

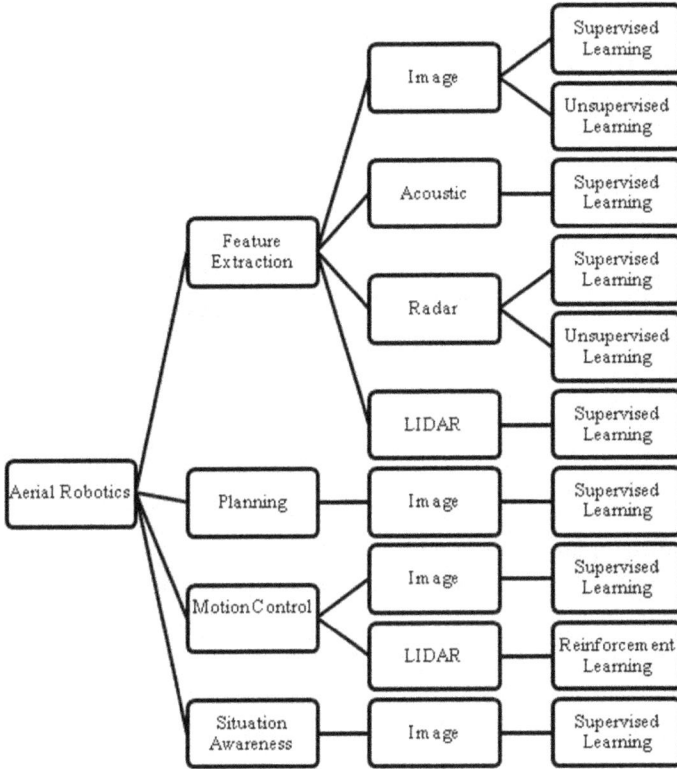

FIGURE 9.6    Summary of deep learning applications in aerial robotics.

Figure 9.6 shows the summary of the deep learning aerial robotics application based on the different tasks performed by the aerial robots.

## 9.8 CONCLUSION

In this chapter, the deep learning applications of aerial robotics were discussed. The architecture of aerial robots was analysed and the different components identified where the deep learning algorithm can be applied based on image and sensor data. The deep learning applications of aerial robots make them more powerful to perform their specific missions. The four different types of deep learning applications of aerial robotics were categorized in this chapter. These applications are: deep learning for feature extraction; deep learning for planning; deep learning for motion control; and deep learning for situation awareness. These applications were summarized in this chapter and all of them are based on learning algorithms. The uses of these deep learning applications of aerial robotics are were discussed for the different application domains such as agriculture, navigation, search and rescue and localization.

## REFERENCES

1. Liew, C. F., DeLatte, D., Takeishi, N., &Yairi, T. (2017). Recent developments in aerial robotics: A survey and prototypes overview. *arXiv preprint arXiv:1711.10085*.
2. Watts, A. C., Ambrosia, V. G., & Hinkley, E. A. (2012). Unmanned aircraft systems in remote sensing and scientific research: Classification and considerations of use. *Remote Sensing, 4*(6), 1671–1692.
3. Gupta, S. G., Ghonge, M. M., & Jawandhiya, P. M. (2013). Review of unmanned aircraft system (UAS). *International Journal of Advanced Research in Computer Engineering &Technology (IJARCET), 2*(4), 1646–1658.
4. Brooke-Holland, L. (2012). Unmanned Aerial Vehicles (drones): An introduction. London: House of Commons Library.
5. Arjomandi, M., Agostino, S., Mammone, M., Nelson, M., & Zhou, T. (2006). Classification of unmanned aerial vehicles. *Report for Mechanical Engineering Class*. Adelaide: University of Adelaide.
6. Cavoukian, A. (2012). *Privacy and drones: Unmanned aerial vehicles* (pp. 1–30). Ontario: Information and Privacy Commissioner of Ontario, Canada.
7. Weibel, R., &Hansman, R. J. (2004, September). Safety considerations for operation of different classes of UAVs in the NAS. In Aiaa 4th Aviation Technology, Integration and Operations (atio) Forum (p. 6244).
8. Authority, C. A. (2004). *Unmanned Aerial Vehicle Operations in UK Airspace: Guidance.*, Directorate of Airspace Policy, 2010, UK.
9. Authority, C. A. (2010). CAP 722 Unmanned aircraft system operations in UK airspace: Guidance. Directorate of Airspace Policy.
10. Homainejad, N., & Rizos, C. (2015). Application of multiple categories of unmanned aircraft systems (UAS) in different airspaces for bushfire monitoring and response. *International Archives of the Photogrammetry, Remote Sensing & Spatial Information Sciences, 40*. pp. 1471–1482.
11. Turanoğuz, E., & Alemdaroğlu, N. (2015, June). Design of a medium range tactical UAV and improvement of its performance by using winglets. In 2015 International Conference on Unmanned Aircraft Systems (ICUAS) (pp. 1074–1083). IEEE.
12. Zakora, B., & Molodchick, A. (2014). *Classification of UAV (Unmanned Aerial Vehicle)*. Retrieved from http://read.meil.pw.pl/abstracts/StudentAbstract_Zakora_Molodchik.pdf.

13. Saari, H., Pellikka, I., Pesonen, L., Tuominen, S., Heikkilä, J., Holmlund, C., ... & Antila, T. (2011, October). Unmanned Aerial Vehicle (UAV) operated spectral camera system for forest and agriculture applications. In *Remote Sensing for Agriculture, Ecosystems, and Hydrology XIII* (Vol. 8174, p. 81740H). International Society for Optics and Photonics.
14. Radoglou-Grammatikis, P., Sarigiannidis, P., Lagkas, T., &Moscholios, I. (2020). A compilation of UAV applications for precision agriculture. *Computer Networks, 172,* 107148.
15. Frachtenberg, E. (2019). Practical drone delivery. *Computer, 52*(12), 53–57.
16. CASA Drone Delivery System, Available at: www.casa.gov.au/drones/industry-initiatives/drone-delivery-systems.
17. Puri, A. (2005). *A survey of unmanned aerial vehicles (UAV) for traffic surveillance.* Department of Computer Science and Engineering, University of South Florida, 1–29.
18. Finn, R. L., & Wright, D. (2012). Unmanned aircraft systems: Surveillance, ethics and privacy in civil applications. *Computer Law & Security Review, 28*(2), 184–194.
19. Erdelj, M., & Natalizio, E. (2016, February). UAV-assisted disaster management: Applications and open issues. In 2016 International Conference on Computing, Networking and Communications (ICNC) (pp. 1–5). IEEE.
20. Estrada, M. A. R. & Ndoma, A. (2019). The uses of unmanned aerial vehicles–UAV's–(or drones) in social logistic: Natural disasters response and humanitarian relief aid. *Procedia Computer Science, 149,* 375–383.
21. Martinez, C., Sampedro, C., Chauhan, A., &Campoy, P. (2014, May). Towards autonomous detection and tracking of electric towers for aerial power line inspection. In 2014 International Conference on Unmanned Aircraft Systems (ICUAS) (pp. 284–295). IEEE.
22. Olivares-Mendez, M. A., Fu, C., Ludivig, P., Bissyandé, T. F., Kannan, S., Zurad, M., ... & Campoy, P. (2015). Towards an autonomous vision-based unmanned aerial system against wildlife poachers. *Sensors, 15*(12), 31362–31391.
23. Carrio, A., Pestana, J., Sanchez-Lopez, J. L., Suarez-Fernandez, R., Campoy, P., Tendero, R., &Marchamalo-Sacristán, M. (2016). UBRISTES: UAV-based building rehabilitation with visible and thermal infrared remote sensing. In Robot 2015: Second Iberian Robotics Conference (pp. 245–256). Cham: Springer.
24. Li, L., Fan, Y., Huang, X., &Tian, L. (2016). Real-time UAV weed scout for selective weed control by adaptive robust control and machine learning algorithm. In 2016 ASABE Annual International Meeting (p. 1). American Society of Agricultural and Biological Engineers.

25. Sanchez-Lopez, J. L., Molina, M., Bavle, H., Sampedro, C., Fernández, R. A. S., & Campoy, P. (2017). A multi-layered component-based approach for the development of aerial robotic systems: The aerostack framework. *Journal of Intelligent & Robotic Systems*, 88(2), 683–709.

26. LeCun, Y., Boser, B., Denker, J., Henderson, D., Howard, R., Hubbard, W., & Jackel, L. (1989). Handwritten digit recognition with a back-propagation network. *Advances in Neural Information Processing Systems*, 2, 396–404.

27. Girshick, R., Donahue, J., Darrell, T., & Malik, J. (2014). Rich feature hierarchies for accurate object detection and semantic segmentation. In Proceedings of the IEEE Conference on Computer Vision and Pattern Recognition (pp. 580–587).

28. Girshick, R. (2015). Fast R-CNN. In Proceedings of the IEEE International Conference on Computer Vision (pp. 1440–1448).

29. Ren, S., He, K., Girshick, R., & Sun, J. (2015). Faster R-CNN: Towards real-time object detection with region proposal networks. *arXiv preprint arXiv:1506.01497*.

30. Lee, J., Wang, J., Crandall, D., Šabanović, S., &Fox, G. (2017, April). Real-time, cloud-based object detection for unmanned aerial vehicles. In 2017 First IEEE International Conference on Robotic Computing (IRC) (pp. 36–43). IEEE.

31. Redmon, J., Divvala, S., Girshick, R., &Farhadi, A. (2016). You only look once: Unified, real-time object detection. In Proceedings of the IEEE Conference on Computer Vision and Pattern Recognition (pp. 779–788).

32. Redmon, J., & Farhadi, A. (2017). YOLO9000: Better, faster, stronger. In Proceedings of the IEEE Conference on Computer Vision and Pattern Recognition (pp. 7263–7271).

33. Liu, W., Anguelov, D., Erhan, D., Szegedy, C., Reed, S., Fu, C. Y., & Berg, A. C. (2016, October). SSD: Single shot multibox detector. In European Conference on Computer Vision (pp. 21–37). Cham: Springer.

34. Ghaderi, A., & Athitsos, V. (2016, December). Selective unsupervised feature learning with convolutional neural network (S-CNN). In 2016 23rd International Conference on Pattern Recognition (ICPR) (pp. 2486–2490). IEEE.

35. Zhou, B., Lapedriza, A., Xiao, J., Torralba, A., &Oliva, A. (2014). Learning deep features for scene recognition using places database.

36. Penatti, O. A., Nogueira, K., & Dos Santos, J. A. (2015). Do deep features generalize from everyday objects to remote sensing and aerial scenes domains? In Proceedings of the IEEE Conference on Computer Vision and Pattern Recognition Workshops (pp. 44–51).

37. Hu, F., Xia, G. S., Hu, J., & Zhang, L. (2015). Transferring deep convolutional neural networks for the scene classification of high-resolution remote sensing imagery. *Remote Sensing*, 7(11), 14680–14707.

38. Gangopadhyay, A., Tripathi, S. M., Jindal, I., &Raman, S. (2015). SA-CNN: Dynamic scene classification using convolutional neural networks. *arXiv preprint arXiv:1502.05243.*

39. Giusti, A., Guzzi, J., Cireşan, D. C., He, F. L., Rodríguez, J. P., Fontana, F., ... & Gambardella, L. M. (2015). A machine learning approach to visual perception of forest trails for mobile robots. *IEEE Robotics and Automation Letters, 1*(2), 661–667.

40. Li, L., Fan, Y., Huang, X., &Tian, L. (2016). Real-time UAV weed scout for selective weed control by adaptive robust control and machine learning algorithm. In 2016 ASABE Annual International Meeting (p. 1). American Society of Agricultural and Biological Engineers.

41. Li, W., Fu, H., Yu, L., & Cracknell, A. (2017). Deep learning based oil palm tree detection and counting for high-resolution remote sensing images. *Remote Sensing, 9*(1), 22.

42. Chen, S. W., Shivakumar, S. S., Dcunha, S., Das, J., Okon, E., Qu, C., ... & Kumar, V. (2017). Counting apples and oranges with deep learning: A data-driven approach. *IEEE Robotics and Automation Letters, 2*(2), 781–788.

43. Hung, C., Xu, Z., &Sukkarieh, S. (2014). Feature learning based approach for weed classification using high resolution aerial images from a digital camera mounted on a UAV. *Remote Sensing, 6*(12), 12037–12054.

44. Rebetez, J., Satizábal, H. F., Mota, M., Noll, D., Büchi, L., Wendling, M., ...& Burgos, S. (2016, April). Augmenting a convolutional neural network with local histograms: A case study in crop classification from high-resolution UAV imagery. In ESANN.

45. Sawarkar, A., Chaudhari, V., Chavan, R., Zope, V., Budale, A., &Kazi, F. (2016). HMD vision-based teleoperating UGV and UAV for hostile environment using deep learning. *arXiv preprint arXiv:1609.04147.*

46. Bejiga, M. B., Zeggada, A., Nouffidj, A., &Melgani, F. (2017). A convolutional neural network approach for assisting avalanche search and rescue operations with UAV imagery. *Remote Sensing, 9*(2), 100.

47. Kim, N. V., & Chervonenkis, M. A. (2015). Situation control of unmanned aerial vehicles for road traffic monitoring. *Modern Applied Science, 9*(5), 1.

48. Kim, H., Kim, D., Jung, S., Koo, J., Shin, J. U., & Myung, H. (2015, October). Development of a UAV-type jellyfish monitoring system using deep learning. In 2015 12th International Conference on Ubiquitous Robots and Ambient Intelligence (URAI) (pp. 495–497). IEEE.

49. The Technion – Israel Institute of Technology, "Technion aerial systems 2016", in Journal Paper for AUVSI Student UAS Competition, 2016.

50. Maturana, D., & Scherer, S. (2015, May). 3D convolutional neural networks for landing zone detection from lidar. In 2015 IEEE International Conference on Robotics and Automation (ICRA) (pp. 3471–3478). IEEE.

51. Mendis, G. J., Randeny, T., Wei, J., & Madanayake, A. (2016, November). Deep learning based doppler radar for micro UAS detection and classification. In MILCOM 2016-2016 IEEE Military Communications Conference (pp. 924–929). IEEE.

52. Jeon, S., Shin, J. W., Lee, Y. J., Kim, W. H., Kwon, Y., & Yang, H. Y. (2017, August). Empirical study of drone sound detection in real-life environment with deep neural networks. In 2017 25th European Signal Processing Conference (EUSIPCO) (pp. 1858–1862). IEEE.

53. Morito, T., Sugiyama, O., Kojima, R., & Nakadai, K. (2016, October). Partially shared deep neural network in sound source separation and identification using a UAV-embedded microphone array. In 2016 IEEE/RSJ International Conference on Intelligent Robots and Systems (IROS) (pp. 1299–1304). IEEE.

54. Delmerico, J., Mueggler, E., Nitsch, J., & Scaramuzza, D. (2017). Active autonomous aerial exploration for ground robot path planning. *IEEE Robotics and Automation Letters*, 2(2), 664–671.

55. Delmerico, J., Giusti, A., Mueggler, E., Gambardella, L. M., &Scaramuzza, D. (2016, October). "On-the-spot training" for terrain classification in autonomous air-ground collaborative teams. In International Symposium on Experimental Robotics (pp. 574–585). Cham: Springer.

56. Kim, D. K., & Chen, T. (2015). Deep neural network for real-time autonomous indoor navigation. *arXiv preprint arXiv:1511.04668*.

57. Sadeghi, F., & Levine, S. (2016). CAD2RL: Real single-image flight without a single real image. *arXiv preprint arXiv:1611.04201*.

58. Shah, U., Khawad, R., & Krishna, K. M. (2016, December). Deepfly: Towards complete autonomous navigation of MAVS with monocular camera. In Proceedings of the Tenth Indian Conference on Computer Vision, Graphics and Image Processing (pp. 1–8).

59. Santana, P., Correia, L., Mendonça, R., Alves, N., & Barata, J. (2013). Tracking natural trails with swarm-based visual saliency. *Journal of Field Robotics*, 30(1), 64–86.

60. Zhang, T., Kahn, G., Levine, S., &Abbeel, P. (2016, May). Learning deep control policies for autonomous aerial vehicles with MPC-guided policy search. In 2016 IEEE International Conference on Robotics and Automation (ICRA) (pp. 528–535). IEEE.

61. Kelchtermans, K., & Tuytelaars, T. (2017). How hard is it to cross the room? Training (recurrent) neural networks to steer a UAV. *arXiv preprint arXiv:1702.07600*.

62. Lin, T. Y., Cui, Y., Belongie, S., & Hays, J. (2015). Learning deep representations for ground-to-aerial geolocalization. In Proceedings

of the IEEE Conference on Computer Vision and Pattern Recognition (pp. 5007–5015).

63. Taisho, T., Enfu, L., Kanji, T., & Naotoshi, S. (2015, December). Mining visual experience for fast cross-view UAV localization. In 2015 IEEE/SICE International Symposium on System Integration (SII) (pp. 375–380). IEEE.

64. Aznar, F., Pujol, M., & Rizo, R. (2016, September). Visual navigation for UAV with map references using convnets. In Conference of the Spanish Association for Artificial Intelligence (pp. 13–22). Cham: Springer.

# CHAPTER 10

# The Memristor and its Implementation in Deep Neural Network Designing

## A Review

Melaku Nigus Getachew, Rashmi Priyadarshini and R.M. Mehra

*Sharda University, Greater Noida, Uttar Pradesh, India,*

*E-mail ids: MelakuNigus@dbu.edu.et; rashmi.priyadarshini@sharda. ac.in; rm.mehra@sharda.ac.in*

## 10.1 INTRODUCTION

This chapter provides a general review of memristor device application in designing hardware deep neural networks such as spiking neural networks (SNNs), multilevel neural networks (MNNs), convolutional neural network (CNNs), and recurrent neural net-works (RNNs). The chapter reviews the network structure, architecture, and algorithm types that are suitable for each kind of deep neural network discussed in this chapter. Deep neural networks, a brain inspired learning model has so far shown an astonishing performance at different kinds of real world applications, from image classification to voice recognition, and natural language processing. However, the conventional computing system architecture, that is to say, Von Neumann computing system architectures, are now suffering from different kindsof emerging technological problems, such as memory

DOI: 10.1201/9781003217237-10

wall due to the growing performance gap between the memory and processor, communication bottleneck due to a bus shared by instruction fetch and data processing operation, and the increasing computing energy demanddue to the end of Dennard scaling. As opposed to this, memristor based deep neural networks that are based on non Von Neumann or parallel processing neuromorphic system architecture are now becoming very promising in the current information processing technologies. For example, memristor crossbar network architectures are now being utilized to perform neural computations that are due to its high density integration natures of the network, due to the possibility of collocating memory and processing in such types of network, their better power efficiency during computations, and the network capability of parallel information processing. In this chapter, design strategies of different types of memristor crossbar based deep neural network are reviewed. All reviewed memristor crossbar deep neural networks are discussed in terms of their network architecture, computing circuit, and the network learning algorithm. Also, the potential application areas of each crossbar deep neural network are discussed.

### 10.1.1 Memeristor

For hardware implementations of neuromophic systems, deep neural networks for that matter, memristor is being becoming the most promising device level component. Many researchers in the field are now becoming very interested in implementing memristor in designing artificial neural networks due to the fact that the device has several promising features, namely: memory effect, nanoscale device size, non-volatility, and passivity. Memristor is the fourth fundamental circuit element theoretically proposed by Chua in 1971 [1], and consequently physically realized by HP Lab research teams in 2008 [2]. The unique and most important characteristic of memristor is that its memristance or device resistance value depends on its historical activity. The main reason why memristor is being used for hardware implementation of biological brain inspired neural network application is its functional resemblance with biological synapse, that is to say the spike time-dependent plasticity (STDP) learning rule can be faithfully implemented in memristor based deep neural network [3]. Another reason to rely on memristor technology for hardware implementation of deep neural network is its crucial role to realize an energy efficient neuromorphic circuitry and also numerous works have been done to investigate the energy efficiency of circuits formed from memristor [4]–[11], and all of them duly reported the benefits of memristor in reducing power dispassion of the neural network. Besides, on the flip-side, it has been widely accepted that hardware implementation of deep neural networks is a natural fit for memristor becausethe deep neural network model fault tolerance nature can provide a means to investigate effects caused by device variations of memristor [12–51]. Depending onthe type and the function of hardware deep neural network, memristor is usually implemented to mimic the functionality of biological synapse in hardware neural circuit [17–50], [52–69].

Sometimes memristor serves as a memory element for the entire neural circuit architecture developed so that the network synaptic weight can be stored into it during the time of computation. In other cases it leverages on the intrinsic plasticity nature of the device component for implementing the Hebian learning in general [70–77] or STDP learning rule in particular [78–112], [260], [113–129]. During the design and development stage of hardware deep neural network in neuromorphic computing systems, memristor is commonly used to form a memristive crossbar in order to mimic the synapses in the neural network [130–232]. Memristor crossbar implementations are widely used because this system architecture is promising to realize a highly dense memristive parallel processing network. Moreover it has been observed that fabricated memristor crossbar networks have performed well. During the training scenes of a memristive crossbar network a zero, negative, and positive synaptic weight programming is required. However thethree types of synaptic weight values cannot be realized from a single memristor implementation as a synapse; for this reason a synaptic circuit formed from multi-memristors has been proposed [233–234], and [291]. Numerous memristive neural network training algorithms have been proposed and all of them were developed by taking the memristor characteristic into account [235–259]. In addition to its utilization in synaptic circuits, memristor have also been utilized in neuron circuit implementation [260–268], [301]. For example, a complex neuronal signal has been generated from memristive neuron circuit proposed in [269–271]. Also memristor is considered as a natural fit to be implemented in stochastic computing system, for instance in [43] a neuron model has been proposed using memristor in order to add stochasticity feature to the model. The classical Hodgkin-Huxley neuronal membrane model hardware implementation has also been accomplished utilizing memristor in [272–273]. According to memristor taxonomy [302–303], there are different types of memristor depending on the types of chemical compound or material used during their physical realizations. For example, redox memristors are physically realized utilizing metal-insulator-metal thin-film structures; these types of devices' electrical conduction process usually depends on the oxidation and redaction process of cations or anions taking place in the device thin-film. There are also devices where conduction process is based on electronic effects that rely on electron trapping and insulator-metal transition process in a Mott insulator [302]. Also, there are phase change memristor devices: their electrical conductance change is usually based on non-crystalline chalcogenide or crystalline chalcogenide thin-film layer microstructure restructuring. In addition to the above

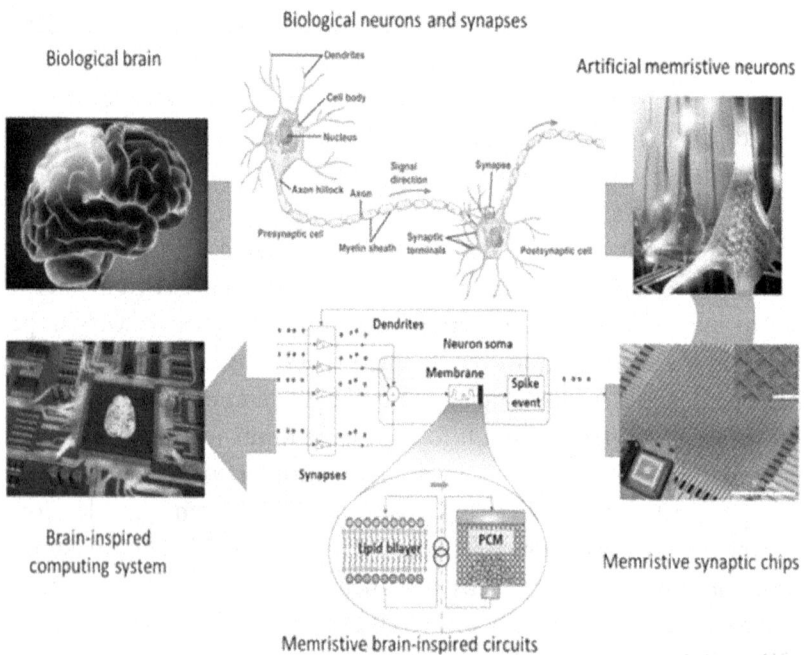

FIGURE 10.1 The figure depicts neuromorphic computing system architecture. The neuromorphic system computing units; namely artificial neurons, and synapses functionalities can be reproduced on memristor based system on-chip. The neuron consists of dendrites, soma, and axon parts. The soma/ cell body is where the nerve membrane occurred and also responsible to generate action potential. The dendrites are where the neuron cell body collects most of its input spikes fired by other neurons.

mentioned types of memristors there are also spin torque transfer and microelectromechanical systems-based memristive devices.

## 10.2 MEMRISTOR IMPLEMENTATION IN BUILDING ARTIFICIAL NEURAL NETWORKS

Computing systems are now becoming sophisticated with very complex task performing capability owing to the advancement made in device level such as the discovery of memristor, ANNs ciruit design archticurelavel, and algorithms development level. The implementations of neural networks in hardware require utilizations of different types of circuits that encompass conventional information processing unit, neural accelerators, and neuromorphic information processing units [274–286]. It's widely accepted that neuromorphic systems are an ideal computing architecture to implement neural networks in hardware, and this neural network implementation in computing systems requires neuromorphic chips, that is to say, neural processing units (NPUs). It's observed from Figure 10.2a that only the number of cores the processing unit possessed, as well as these computing cores organization used by each type of processing units, that accounts for the existing difference between central processing units (CPUs), graphics processing units (GPUs), and NPUs. The CPUs contains at most several tens of computingcores, whereas both GPUs and NPUs contains more than several thousand [287]. Despite their similarity in using many computing cores for parallel processing both NPUs and GPUs are different from each other in operation and system organization. The GPUs' computing cores usually serves as a computing element rather than asmemory elements, although there is static random access memory (SRAM) in GPUs' architecture but they are there to be used as a cache memory. This means memory and processing are notin the same place and/or collocated in GPUs' circuit architecture just like as is the case in CPUs. However, NPUs' computing cores have their own memory unitsor synapses. Different from the architecture of CPUs and GPUs, NPUs' memory and processing are collocated, that is to say, that just like in a biological brain, in-memory computation is possible in NPUs. Nevertheless, currently NPUs coresutilize volatile devices such as SRAM or dynamic random access memoryas synapses [283,288,289]. Even thoughprocessing and memory operation are collocated in NPU cores, non-volatile external storage is required for synaptic weight storage functionality. If we use memristor devices as synapses, there is no external storage required and therefore realizationof in-memory computing is also

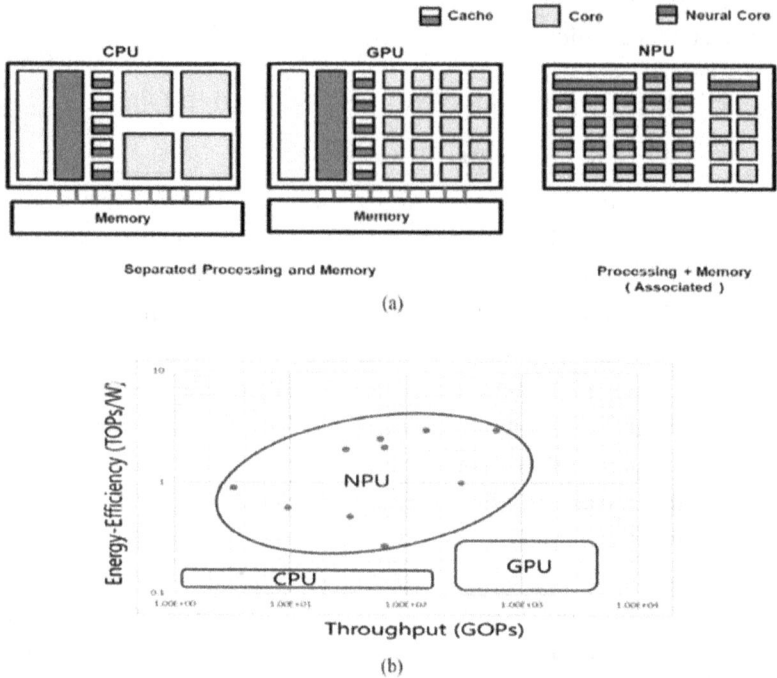

FIGURE 10.2    Both CPUs and GPUs are conventional processing units, however in both processing units the numbers, and organizations of computing cores are different. In-memory computing is possible in NPUs, but in conventional processing units memory and processing are separated, b. the energy efficiency performance of the deep neural network processors, NPUs is more power efficient than CPUs and GPU.

possible [290]. Figure 10.2b shows a comparison of the computing energy efficiency among CPUs, GPUs, and NPUs. The energy efficiency increases in the order of CPUs, GPUs, and NPUs.

## 10.3  DEEP LEARNING

In deep learning, multilayer computational models are learning representation of data with multiple levels of abstraction. This kind of deep learning computational models are now enhancing the technology implemented in different practical applications such as speech recognition, visual object recognition, and object detection, as well as in pharmaceutical drug discovery and genomics. Deep learning usually processes large data sets by implementing an algorithms such as backpropagation. With the help of this backpropagation algorithm, the synaptic weight in each layer of the network changes its value so that the data representation in each layer of the network can be computed from the previous layer representation. There are SNN, MNN, CNN, and RNN structure in deep learning; nevertheless, CNN is the most well celebrated for processing image, video, and audio datasets such as text and speech. Deep learning brought a breakthrough by solving numerous problems in artificial intelligence that had proved a major challenge for decades for the research community in the field. The main power of deep learning relies in its ability to find patterns in a very high-dimensional data set at a very highspeed. In addition to its application in different domains such as science, business, and government, deep learning will bring us additional breakthroughs in the near future, due to the fact that this technology requires very little hand engineering since it has the potential to take advantage of the available computation and data set size to learn as the net computing capability powered by newly developing architectures and algorithms.

## 10.4 MEMRISTOR CROSSBAR ARCHITECTURE

Memristor crossbar architecture is a computing-in-memory architecture which has both high density and power efficiency features. The crossbar

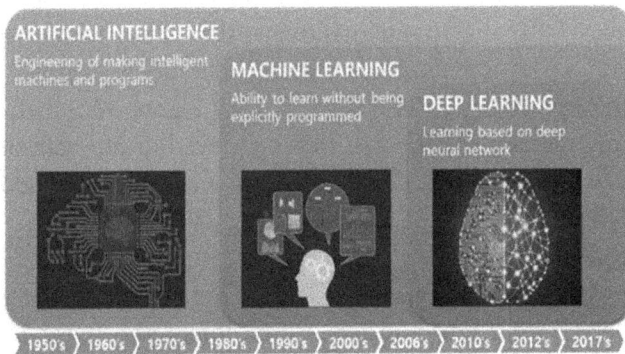

FIGURE 10.3    AI, machine learning, and deep learning evolution through time. The relationship between AI, machine learning and deep learning can be defined as that AI is broad topic that includes all types of intelligent computing; namely machine learning and deep learning. Machine learning is another types of intelligence computing that makes a computing machine learn from date it feeds rather than programming as it was traditionally. Deep learning is a type of intelligent computing implemented to train machine to be smarter by mainly using multiple layers of neural network technologies. https://blogs.nvidia.com/blog/2016/07/29/whatsdi_erence-arti_cial-intelligencemachine- learning-deep-learning-ai.

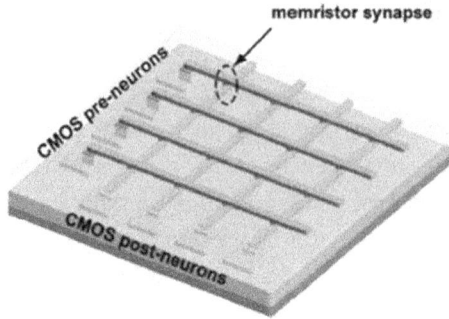

FIGURE 10.4   Crossbar architecture.

formed from a couple of perpendicular nanowire layers, and the upper
layer acted as a top electrode, also at these wires cross joint a memristor
device inserted so that it acts as a synapse in the entire architecture
(see Figure 10.4). There are many neural networks, such as spike neural
networks, multilayer neural networks, convolutional neural networks, and
recurrent neural networks, which are now being realized with the help of
the memristor crossbar architecture. This large scale implementation of
crossbar architecture is powered by its promising features such as high-
density architecture due to the memristor nanoscale size, and power efficiency of the architecture due to the existence of memory and processing at
the same place [291, 292]. In memristor crossbar neural networks there are
three network operations available: reading operations, writing operations,
and training operations. It is observed from Figure 10.5 that, there are
two types of common biasing methods employed to program or read the
nanoscale memristor device sandwiched between word lines and bit lines,
namely the $\frac{V}{2}$, and $\frac{V}{3}$ bias schemes. When the $\frac{V}{2}$ bias methods employed
the higher voltage V, and ground voltage 0 are applied to the exact pos-
ition of selected word line, and selected bit line, respectively. During this
time a half voltage $(\frac{V}{2})$ is connected to other word and bit lines in the
crossbar, therefore as a results of these three types of applied bias voltages
that are V,$\frac{V}{2}$, and 0, the selected memristor in the crossbar would become
under V, the half selected would become under $(\frac{V}{2})$ bias voltage, and that
of the unselected one would become under voltage ground bias. Once the

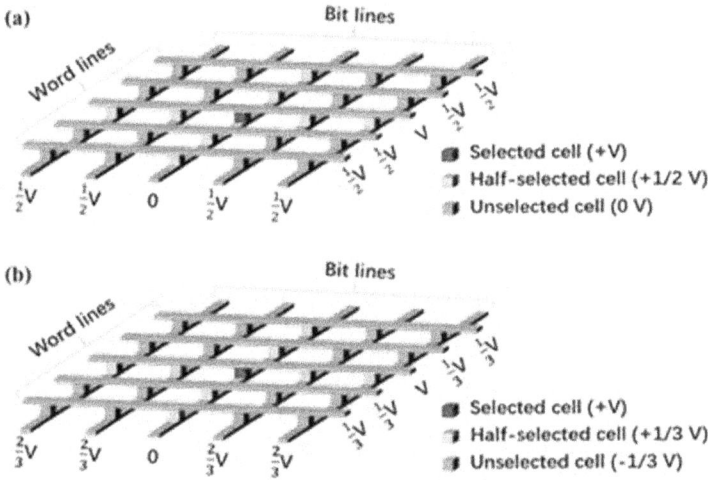

FIGURE 10.5   Crossbar architecture that shows the reading, writing and training operation of a network.

memristor crossbar properly biased in this way then the conductance of the particular memristor element that is biased under the full V circuit voltage can be calculated, hence the reading operation is performed. On the other hand, when the $\frac{V}{3}$ bias method is employed, higher voltage V, and ground voltage 0 are applied to the exact position of selected word line and selected bit line, respectively, however $\frac{V}{3}$ is connected to unselected word lines, and that of $2\frac{V}{3}$ bias connected to unselected bit lines. As a result the selected, half selected, and unselected memristors in the crossbar would become under full V, $\frac{V}{3}$, and $\frac{-V}{3}$ bias, hence in this way the memristor crossbar reading, writing, and training operations can be accomplished [304].

## 10.5  SPIKE NEURAL NETWORKS

SNNs are implemented in hardware in different forms. These implementations use different kinds of neuron models depending on the types of application for which the neural network is intended to be used. For example, if one has plan to form SNN network for direct investigations of neuroscience problem in silico test bed, then rather than bi-dimentional generalized adaptive integrate and fire nuron models, a more biologically- plausiable membrane ion-channel conductance-based neuron circuit models have to be used to design a hardware SNN network or neuromorphic system. Similar to biologicalneural networks, in spike neural networks, spikes are implemented for informationprocessing. In SNNs' dynamics modeling it is not only the neurons' state and weight that are considered but also the spike timing between the pre-and postsynaptic neuron firing. Out of several artificial neural networks, or deep neural networks forthat matter, the spiking neural network is the one which is built by mimicking thenetwork of the biological brain where nerve cells are transmitting information usingspikes and making connection with each other via synapse. For the hardware realization of a working SNN, memristor deviance is a promising candidate for itsimplementation as a synapse in the network. In SNN the weight of a synapse inthe network is updated using the STDP learning rule; by using this learning rule both unsupervised and supervised, machine learning methods can be demonstrated more efficiently [300]. SNN is now being used for numerous applications in different field. The SNN circuit is very complex and itspractical implementation requires further simplification of the network structureby taking various factors into considerations, such as the network area occupation, and power efficiency. A memristive deep neural network can be described from the perspective of the algorithm type the hardware neural network used during the traning cycle, from the perspective of the network architecture used during the design stage, and from that of the network circuit level implementation.

In Figure 10.6, a three layer spiking neural network is depicted in order to show the network architecture. This network is a feed-forward network.

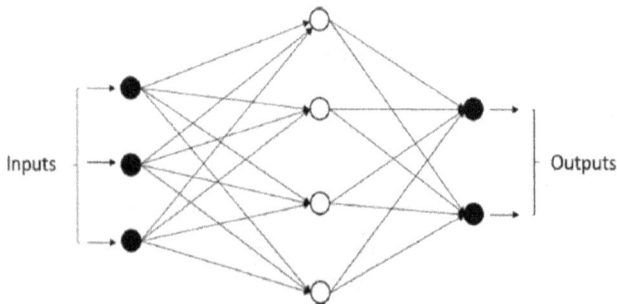

FIGURE 10.6   The three layer spike neural network architecture.

In this SNN, neurons can be modeled using different kinds of modeling types depending on the type of application the network is developed. The available neuron models span from biologically plausible to more mathematically simplified behavioral neuron models. The neuronal spike transmission process takes placebetween the presynaptic neuron and/or spike sending neurons and postsynapticneuron and/or spike receiving neuron, also the medium these two types of neuron used to communicate is called a synapse. Assume there are N number of connected synapses to send spikes for the input neuron, and also consider that the $i^{th}$ synapse $i = i...N$ transmits $G_i$ spike to the input neuron, and in the incoming spike train each spike arrives to the input neuron at a specified time represented as $g_i = t_i^1, t_i^2,$

$... t_i^g$. The latest neuron spike firing time denoted as $t^{fr}$, therefore the synaptic neuron internal state change can be expressed as:

$$u(t) = \sum_{i=1}^{N} \sum t_i^g \epsilon \, g_i \omega_i \epsilon \left(t - t_i^g\right) + \eta\left(t - t_i^{fr}\right), t_i^g > t_i^{fr} \qquad 1$$

Where $\omega_i$, and $\epsilon\left(t\right)$, and the $i^{th}$ synaptic weight, and spike response function, it is important to bear in mind that the model that is describing the postsynaptic neuron can be extended by takingthe network architecture into account so as to describe the entire SNN dynamics. The SNN is usually implemented as an algorithm called a spike time dependent plasticity (STDP) mechanism (see Figure 10.7). The STDP learning is an unsupervised learning algorithm based on dependencies between presynaptic, and postsynaptic spikes. In a given synapse, when a postsynaptic spike occurs in a specific time window after a presynaptic spike, the weight of this synapse is increased. If the postsynaptic spike appears before the presynaptic spike,

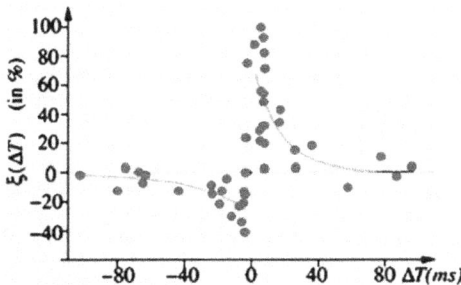

FIGURE 10.7   STDP curve used for learning which is experimentally recorded from a biological synapse.

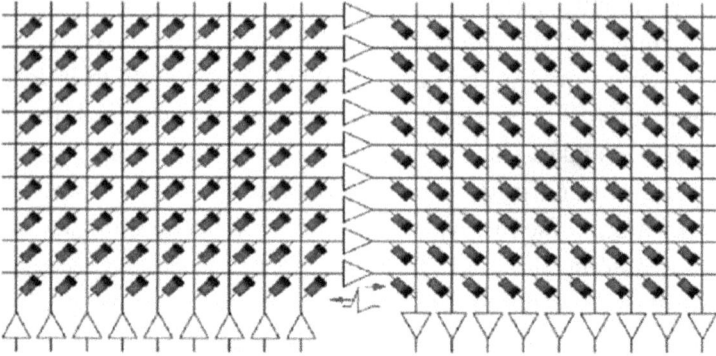

FIGURE 10.8    Hybrid CMOS-memristor implementations of SNN circuit [304].

a decrease in the weight occurs assuming that inverse dependency exists between pre- and postsynaptic spikes. The strength of the weight change is a function of time between presynaptic and postsynaptic spike events.

In Figure 10.8 an SNN circuit level implementation is depicted. This SNNis a three layer deep neural network with memristor synapses. In this particular network the neurons are based on complementary metal-oxide semiconductor (CMOS) device technology whereas that of the synapses are memristors, therefore such kinds of neural network implementations are called hybrid CMOS-memristor neuromorphic circuits.

## 10.6 FEED-FORWARD NEURAL NETWORKS

There are always different types of neural network connectivity typologies thatmight be required for neuromorphic computing system hardware circuit implementation:feed-forward; feed-forward with some recurrence; sparsely connected recurrent; local-connected recurrent; and fullyconnected recurrent [305–309]. Therefore, when designing a deep neural network in hardware, there are two mainchallenging issues: determining the level of connectivity between neuronsand layers of neurons in the net; and finding the hardware that is capable of accommodating the desired connectivity in the net. Multi-layer perceptron, and also, with the rise of deep learning, CNN are both feed-forward neural networks.

$$STDP\left(\Delta t\right)=\varsigma\left(\Delta T\right)=\{A^{-}exp\,exp\left(\frac{\Delta t}{\tau^{-}}\right),if\ \sim\sim\Delta t\leq-20,$$

$$if\ \sim\sim-2<\Delta t\leq2A^{+}\,exp\,exp\left(\frac{\Delta t}{\tau^{+}}\right),if\ \sim\sim\Delta t\geq2 \qquad 2$$

## 10.7 MULTILEVEL NEURAL NETWORKS

The perceptron, that is to say, a single-layer neural network only classifies linear data; however in order to classify linearly indivisible data we need to use multilayer perceptron or MNNs. The multilayer neuralnetwork is a quintessential deep neural network, and this neural network is capable of accomplishing a complex learning process if there is available massive computational power for doing that [293,294]. However this complex learning process needs implemented local rules such as back propagation algorithms based on large numbers of the network synaptic weights updates [295,296]. The multilayer neuralnetwork architecture, specifically a double layer deep neural network is shown in Figure 10.9, and historically this type of network evolved from single-layer perceptron. In this type of neural network architecture, all neurons placed in the lowerand upper layers are completely connected, but neuron connection in the samelayer and cross layer neuron connections are both prohibited, as depicted in Figure 10.9. As observed from the network architecture in Figure 10.9, the MNN is formed from an input layer, one hidden layer, and an output layer, hence MNN is a quintessential deep neural network.

The network variables X, W, and Y in Figure 10.9 represent the inputsignal, synaptic weight values, and an output value, respectively. For example, consider the double layer deep neural network shown above in Figure 10.9, and assume this network contains N numbers of input neurons, and their corresponding input signal then can be represented as $x_1, \ldots x_j, x_N$, the network bias given by $b_j$, the network activation function is f, therefore the input signal effect at the hidden layer can be computed as $N_{11} = f(x_1)$ $w_{11} + x_2 w_{21} + b, N_{12} = f(x_1) w_{21} + x_2 w_{22} + b$, and the network output also $Y_1 = f(N_{11})w_{11} + N_{12}w_{21} + b$. In Figure 10.10, a memristor-based circuit diagram of a double layer MNN hardware implementation is presented.

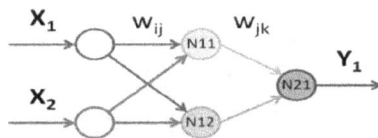

FIGURE 10.9   Logic scheme of the implemented neural network with two inputs, two hidden and one output neurons [304].

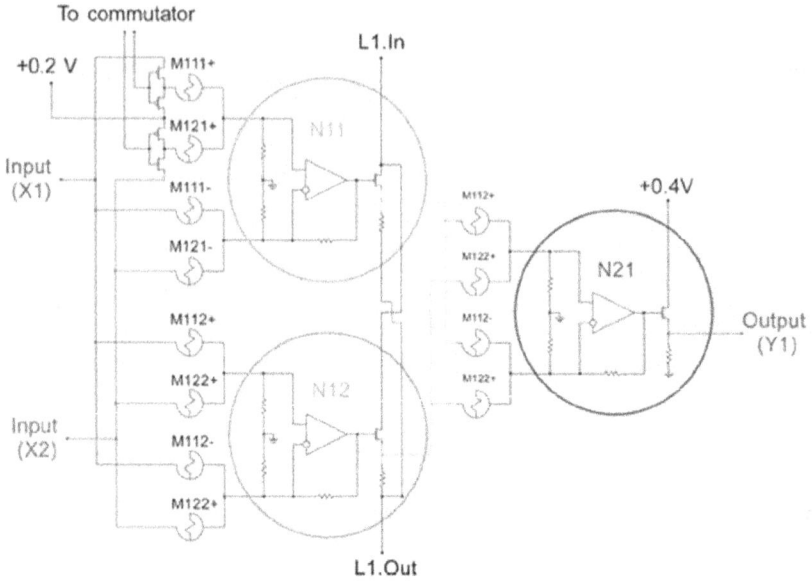

FIGURE 10.10    Memristor based two-layer multilayer neural network [304].

## 10.8 CONVOLUTIONAL NEURAL NETWORKS

The type of deep neural network that is implemented in feed-forward network architecture, which is also simple to train, and better than networks where theirlayers are fully connected, is called convolutional neural network (CNN) (see Figure 10.11). CNNs are designed to process input data which is providedas a multiple array data set. For example, a color image can be representedas a multiple array data set or as a three 2D data array where each array, inturn, represents the image pixel intensities. Therefore, if the net input data set issignals and sequences, including language, then such types of input are represented by 1D multiple arrays; when the inputs are in image or audio spectrogram formit they are represented by 2D multiple arrays; whereas a video or volumetric image is represented by 3D multiple arrays. Based on the properties of natural signals behind every CNN there are four essential working ideas: local connections; shared weight; pooling; and the use of many layers in the network. From the architectural perspective of a typical CNN, the first two stages of the nets constitutes of convolutional layers and pooling layers. The role of CNN convolutional layers are converting theinput data set into its equivalent more abstract level of representation, therefore these layers functionin a similar way to that of the hidden layers of an ordinary neuralnetwork. The computation

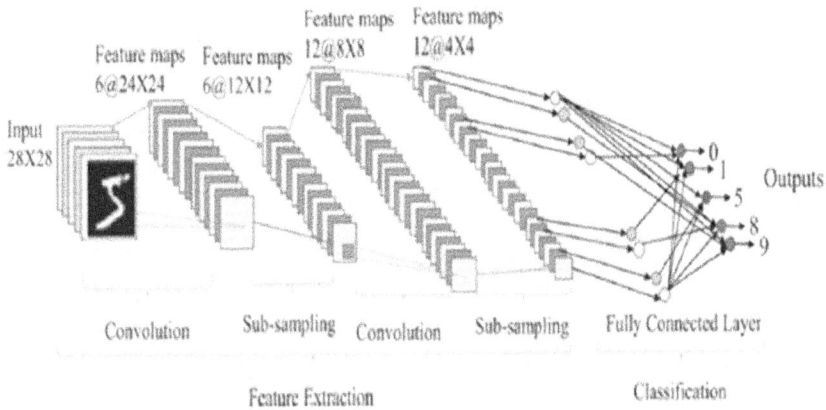

FIGURE 10.11    Block diagrams of convolutional neural network [304].

between the inputs and hidden layer neurons is performed with the help of local connectivity rather than full connectivity schema. There is at least one kernel in a conventional layer, and it can serve as a future extractorand is capable of sharing its weight with all neurons. Both the shared weight and thelocal connectivity are essential for reducing the total number of parameters. On the other hand, the pooling layer is responsible at reducing the input dimension and to carry out this task a pre-specified pooling method, such as averaging pooling, min-pooling, fractional max-pooling, and stochastic pooling, needs to be implemented.

## 10.9  RECURRENT NEURAL NETWORKS

RNNs are neural network types that usually allow for cycles in the network. For example, Hopfield neural network is a type of RNN, which is also common in RNN implementations. RNNs specialize in sequential data processing such as voice recognition [297]. Compared with multilayer networks, in RNNs synaptic weight sharing is possible between neurons [297,298]. And, others in RNNs, lengths history represented by neurons with recurrent connections. Besides, recurrent networks can learn to compress whole histories in low dimensional space; while feed-forward networks compress (project) just single words, recurrent networks have the possibility to form short term memory, so they can better deal with position invariance [299] RNN architecture. In Figure 10.12 a simplified RNN architecture is presented [297]. The RNN circuitis presented to the left side of Figure 10.12 and this network weight matrices denoted by U,V, and W, each corresponding to input-to-hidden layer connection, hidden-to-output

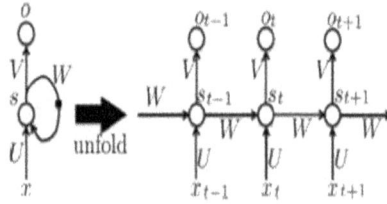

FIGURE 10.12    Architecture of RNN [304].

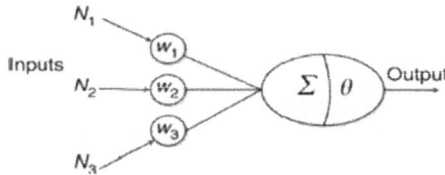

FIGURE 10.13    The three-input binary artificial neuron [304].

layer connection, and hidden-to-hidden layer connection, respectively. Also the vector of activations in the circuit is represented by circles. The right of Figure 10.12 is a time-unfolded own graph, where each node is now associated withone particular time instance.

## 10.10 HOPFIELD NEURAL NETWORKS

Memristor-based Hopfield networks (MHNs) are apreferable model for memristivecircuit network where usually there is complex switching phenomena [300]. As depicted in Figure 10.13 a neuron receives input from three sources, and each input, Ni = (i = 1; 2; and3), is connected to a synapse with synaptic weight of $w_i$. Therefore, the three-input binary neuron output can be given as:

$$Y = sign\left(\sum_{i=1}^{3} w_i N_i - \vartheta\right)$$

Where Y, $\vartheta$ are the neuron output and the threshold function respectively. The 3-bit memristor-based neuron circuit is depicted in Figure 10.14. All the inputs inthe circuit summed up using the op-amp. The positive, negative, and zero synaptic weights are obtained by controlling the circuit switches $s_1; s_2$ and $s_3$. Also, toreform the signals transmission gates labeled by $B_1, B_2$, and $B_3$ ar utilized, and thistask performed without changing the polarity of the signal. However, the negative synaptic weight is obtained

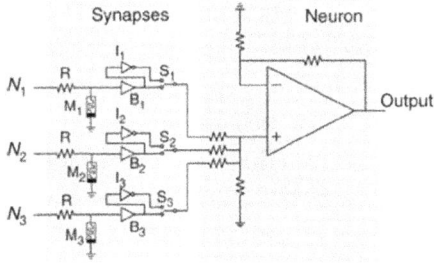

FIGURE 10.14    The 3-bit memristor-based neuron circuit implementation [304].

by invertors $I_1, I_2$, and $I_3$. In this particular architecture $w_{i,j}$ represents the weight of a synaptic connection from neuron i to neuron j. This synaptic weight is mapped to resistance of the corresponding memristor $M_{i,j}$.

$$M = \left( M_{11} M_{12} M_{13} M_{21} M_{22} M_{23} M_{31} M_{32} M_{33} \right),$$
$$W = \left( W_{11} W_{12} W_{13} W_{21} W_{22} W_{23} W_{31} W_{32} W_{33} \right)$$

## 10.11  NEURAL NETWORK ALGORITHMS AND LEARNING

Similar to the learning capability of the biological brain, the artificial neural network has an amazing potential in adapting its surrounding environment in order to accomplish certain tasks. Synaptic plasticity phenomenon, which is a synaptic connection strength change occurring in the biological neural network, is responsible for biological brain learning capability. In the artificial neural network this synaptic plasticity is mimicked by utilizing memristor as a synapse; as a result several learning algorithms were developed, for the artificial neural network to learn.

The algorithm types selected to implement in to a particular type of network was also dictated by the type of basic components or computing units of the neuralnetwork such as neuron, synapse models as well as the model of the network itself [309]. Whenever, we think of the algorithms for neuromorphic systems there area variety of issues to consider. For example, the synapse, neurons, andneural network models choice can affect the type of algorithms that are dependent on the model of neurons and networks as well as the topology of the network. The second case that needs to be considered during algorithm development is whetherthe training or learning should take place on-chip or unsupervised, that is to say, on-line algorithm or off-line algorithm types, respectively.

The main reasons that make neuromorphic systems a promising candidate fornew computing technologies in post-Moore's law era is their system architecture potential for on-line (usually unsupervised) on-chip learning. Back-propagation is the most well-known algorithm type for programming aneuromorphic systems. This back-propagation algorithm is a type of supervised learning. The feed-forward neural networks, recurrent neural networks, and convolutional neural networks are usually trained using back-propagation algorithm learning methods. However, implementing back-propagation algorithms on hardware is a bit difficult, since there are a lot of factors that cause this, such as this algorithm isusually restrictive on the type of neuron and network models as well as the type of topology; the network has, for example, feed-forward, feed-forward with some recurrence, sparsely connected recurrent, locally-connected recurrent, and fullyconnected recurrent typologies. On the other hand, there are also unsupervised learning rules for implementation in neuromorphic systems. This kind of on-chip, on-line self-learning mechanism is fundamental to see the full potential advantages of neuromorphic systems. Hebbian type learning rules are the most common types of on-line, unsupervised learning mechanism in neuromorphic systems. The STDP learning mechanism is that, depending on the spike firing time of the pre-and postsynaptic neurons, thesynaptic weight value will change accordingly. For example, if the presynaptic neurons fire, synaptic weight will increase, whereas the weight will decrease if the postsynaptic fires before the pre one.

Generally, in the case of supervised learning, the given dataset is divided into input dataset and desired output dataset. As such, in this case the neural network is trained so as to give a result that is similar to the desired one. In the case of unsupervised learning, there is one type of dataset, therefore the network in this case is trained so as to find a specific patter within the given dataset, such that future input dataset of the network identified according to the recorded patterns from the previous one. However, in the case of reinforcement learning there are certain actions, rewards, and punishments in response to the type of action the neural network performs. This learning method is similar to the learning of animals in nature.

## 10.12 HARDWARE IMPLEMENTATION OF DEEP NEURAL NETWORK APPLICATION

The realizations of deep neural network in hardware require an effective model selection for neurons and synapses; at the circuit level it require the introduction of a novel device technology, as well as an efficient algorithm. The hardware deep neural network full potential application is not realized yet, though the discovery of memristor has now accelerated the developments of such kinds of technologies. There are several technological artificial intelligence problems, such as machine translation, intelligent question-and-answer, and game play, which are solved by hardware deep neural network implementations that are, in one way or another, have been developed utilizing memristor devices technology. The potential application area and prospects of memristor based hardware deep neural network is very broad, encompassing brain-inspired computing, machine-brain interface technologies, and most all kinds of artificial intelligence application areas.

## 10.13 CONCLUSION

In this chapter several kinds of memristive hardware deep neural networks were presented and also their corresponding circuit level implementations were discussed. The memristor device role in realizing neuron circuit, synapse circuit, and deep neural network circuits were discussed. The memristive hardware deep neural networks, such as SNN, MNN, CNN, and RNN circuit level implementations, are thoroughly discussed. When we see the number of publications published, and experimental works done by both academia and industry that are related to memristors, and also related to the memristor implementations for artificial intelligence technology developments, there is no doubt that memristor is indeed an ideal device for realizing effective future hardware neural networks.

## REFERENCES

1.  L. Chua, Memristor: the missing circuit element, *IEEE Transactions on Circuit Theory*, vol. 18, no. 5, pp. 507–519, 1971.
2.  B. Linares-Barranco and T. Serrano-Gotarredona, Memristance can explain spike-timedependent-plasticity in neural synapses, *Nature Proceedings*, vol. 1, p. 2009, 2009.
3.  L. Deng, D. Wang, Z. Zhang, P. Tang, G. Li, and J. Pei, Energy consumption analysis for various memristive networks under different learning strategies, *Physics Letters A*, vol. 380, no. 7, pp. 903–909, 2016.
4.  C.-C. Hsieh, A. Roy, Y.-F. Chang, D. Shahrjerdi, and S. K.Banerjee, A sub-1-volt analog metal oxide memristive-based synaptic device with large conductance change for energy-efficient spike-based computing systems, *Applied Physics Letters*, vol. 109, no. 22, p. 223501, 2016.

5. B. Liu, M. Hu, H. Li, Y. Chen, and C. J.Xue, Bio-inspired ultra-lower-power neuromorphic computing engine for embedded systems, in Proceedings of the Ninth IEEE/ACM/IFIP International Conference on Hardware/ Software Codesign and System Synthesis. IEEE Press, 2013, pp. 23:1–23:1.

6. B. Rajendran, Y. Liu, J.-s. Seo, K. Gopalakrishnan, L. Chang, D. J. Friedman, and M. B. Ritter, Specifications of nanoscale devices and circuits for neuromorphic computational systems, *IEEE Transactions on Electron Devices*, vol. 60, no. 1, pp. 246–253, 2013.

7. T. M.Taha, R. Hasan, C. Yakopcic, and M. R.McLean, Exploring the design space of specialized multicore neural processors, in Neural Networks (IJCNN), The 2013 International Joint Conference on. IEEE, 2013, pp. 1–8.

8. T. M. Taha, R. Hasan, C. Yakopcic, and M. R. McLean, Low power neuromorphic architectures to enable pervasive deployment of intrusion detection systems, in R. E. Pino, A. Kott, and M. Shevenell (eds), *Cybersecurity Systems for Human Cognition Augmentation*. Springer, 2014, pp. 151–168.

9. T. Tang, R. Luo, B. Li, H. Li, Y. Wang, and H. Yang, Energy efficient spiking neural network design with RRAM devices, in 2014 14th International Symposium onIntegrated Circuits (ISIC). IEEE, 2014, pp. 268–271.

10. Y. Wang, T. Tang, L. Xia, B. Li, P. Gu, H. Yang, H. Li, and Y. Xie, Energy efficient RRAM spiking neural network for real time classification, in Proceedings of the 25th edition on Great Lakes Symposium on VLSI. ACM, 2015, pp. 189–194.

11. L. Chen, C. Li, T. Huang, and X. Wang, Quick noise-tolerant learning in a multi-layer memristive neural network, *Neurocomputing*, vol. 129, pp. 122–126, 2014.

12. B. Gao, P. Huang, B. Chen, L. Liu, X. Liu, and J. Kang, Origin and suppressing methodology of intrinsic variations in metal-oxide RRAM based synaptic devices, in, 12th IEEE International Conference onSolid-State and Integrated Circuit Technology (ICSICT). IEEE, 2014, pp. 1–3.

13. D. Querlioz, O. Bichler, and C. Gamrat, Simulation of a memristorbased spiking neural network immune to device variations, in 2011 International Joint Conference on Neural Networks (IJCNN). IEEE, 2011, pp. 1775–1781.

14. D. Querlioz, O. Bichler, P. Dollfus, and C. Gamrat, Immunity to device variations in a spiking neural network with memristive nanodevices, *IEEE Transactions on Nanotechnology*, vol. 12, no. 3, pp. 288–295, 2013.

15. X. Zhu, Impact of imprecise programming of memristor on building hardware neural network, in International Conference on Electrical and Control Engineering (ICECE). IEEE, 2011, pp. 4527–4529.

16. J. Burger and C. Teuscher, Volatile memristive devices as short-term memory in a neuromorphic learning architecture, in Proceedings of the 2014 IEEE/ACM International Symposium on Nanoscale Architectures. ACM, 2014, pp. 104–109.

17. Z. Dong, S. Duan, X. Hu, L. Wang, and H. Li, A novel memristive multi-layer feedforward small-world neural network with its applications in PID control, *The Scientific World Journal*, vol. 2014, 2014.

18. A. Emelyanov, V. Demin, D. Lapkin, V. Erokhin, S. Battistoni, G. Baldi, S. Iannotta, P. Kashkarov, and M. Kovalchuk, PANI-based neuromorphic networks– firstresults and close perspectives, in 2015 International Conference on Memristive Systems (MEMRISYS). IEEE, 2015, pp. 1–2.

19. A. Emelyanov, D. Lapkin, V. Demin, V. Erokhin, S. Battistoni, G. Baldi, A. Dimonte, A. Korovin, S. Iannotta, P. Kashkarov et al., First steps towards the realization of a double layer perceptron based on organic memristive devices, *AIP Advances*, vol. 6, no. 11, p.111301, 2016.

20. H. Manem, K. Beckmann, M. Xu, R. Carroll, R. Geer, and N. C. Cady, An extendable multi-purpose 3D neuromorphic fabric using nanoscale memristors, in 2015 IEEE Symposium on Computational Intelligence for Security and Defense Applications (CISDA). IEEE, 2015, pp. 1–8.

21. C. Merkel, D. Kudithipudi, and R. Ptucha, Heterogeneous CMOS/memristor hardware neural networks for real-time target classification, in SPIE Sensing Technology+ Applications. *International Society for Optics and Photonics*, 2014, pp. 911–908.

22. O. Such, M. Klimo, and O. Skvarek, Phoneme discrimination using a pair of neurons built from CRS fuzzy logic gates, in Proceedings of the International Conference on Numerical Analysis and applied mathematics 2014 (ICNAAM-2014), vol. 1648. AIP Publishing, 2015, p. 280010.

23. C. Merkel and D. Kudithipudi, Neuromemristive extreme learning machines for pattern classification, in 2014 IEEE Computer Society Annual Symposium onVLSI (ISVLSI). IEEE, 2014, pp. 77–82.

24. C. Merkel and D. Kudithipudi, A current-mode CMOS/memristor hybrid implementation of an extreme learning machine, in Proceedings of the 24th Edition of the Great Lakes Symposium on VLSI. ACM, 2014, pp. 241–242.

25. M. Suri and V. Parmar, Exploiting intrinsic variability of filamentary resistive memory for extreme learning machine architectures, *IEEE Transactions on Nanotechnology*, vol. 14, no. 6, pp. 963–968, 2015.

26. A. Serb, J. Bill, A. Khiat, R. Berdan, R. Legenstein, and T. Prodromakis, Unsupervised learning in probabilistic neural networks with multi-state metal-oxide memristive synapses, *Nature Communications*, vol. 7, p. 12611, 2016.

27. D. Garbin, O. Bichler, E. Vianello, Q. Rafhay, C. Gamrat, L. Perniola, G. Ghibaudo, and B. DeSalvo, Variability-tolerant convolutional neural network for pattern recognition applications based on oxRAM synapses, in, 2014 IEEE InternationalElectron Devices Meeting (IEDM). IEEE, 2014, pp. 2841–2844.

28. D. Garbin, E. Vianello, O. Bichler, Q. Rafhay, C. Gamrat, G. Ghibaudo, B. DeSalvo, and L. Perniola, HfO2-Based OxRAM Devices as Synapses for Convolutional Neural Networks. IEEE Transactions on Electron Devices, Institute of Electrical and Electronics Engineers, vol. 62, no. 8, pp. 2494–2501, 2015.

29. D. Garbin, E. Vianello, O. Bichler, M. Azzaz, Q. Rafhay, P. Candelier, C. Gamrat, G. Ghibaudo, B. DeSalvo, and L. Perniola, On the impact of OxRAM-based synapses variability on convolutional neural networks performance, in 2015 IEEE/ACM International Symposium onNanoscale Architectures (NANOARCH). IEEE, 2015, pp. 193–198.

30. D. Kudithipudi, Q. Saleh, C. Merkel, J. Thesing, and B. Wysocki, Design and analysis of a neuromemristive reservoir computing architecture for biosignal processing, *Frontiers in Neuroscience*, vol. 9, p. 502, 2015.

31. S. Duan, Z. Dong, X. Hu, L. Wang, and H. Li, Small-world Hopfield neural networks with weight salience priority and memristor synapses for digit recognition, *Neural Computing and Applications*, pp. 1–8, 2015.

32. X. Guo, F. Merrikh-Bayat, L. Gao, B. D.Hoskins, F. Alibart, B. Linares-Barranco, L. Theogarajan, C. Teuscher, and D. B.Strukov, Modeling and experimental demonstration of a Hopfield network analog-to-digital converter with hybrid CMOS/memristor circuits, *Frontiers in Neuroscience*, vol. 9, 2015.

33. S. Hu, Y. Liu, Z. Liu, T. Chen, J. Wang, Q. Yu, L. Deng, Y. Yin, and S. Hosaka, Associative memory realized by a reconfigurable memristive Hopfield neural network, *Nature Communications*, vol. 6, pp. 1–8, 2015.

34. B. Liu, Y. Chen, B. Wysocki, and T. Huang, The circuit realization of a neuromorphic computing system with memristor-based synapse design, in T. Huang, Z.Zeng, C. Li, and C. S. Leung, *Neural Information Processing*. Springer, 2012, pp. 357–365.

35. Reconfigurable neuromorphic computing system with memristor-based synapse design, *Neural Processing Letters*, vol. 41, no. 2, pp. 159–167, 2015.

36. A., Khodamoradi, Reshaping Deep Neural Networks for Efficient Hardware Inference. University of California, San Diego, 2021. https://escholarship.org/uc/item/6m01b8sm.

37. Y. V.Pershin and M. DiVentra, Experimental demonstration of associative memory with memristive neural networks, *Neural Networks*, vol. 23, no. 7, pp. 881–886, 2010.

38. L. Wang, H. Li, S. Duan, T. Huang, and H. Wang, Pavlov associative memory in a memristive neural network and its circuit implementation, *Neurocomputing*, vol. e 171, Issue C, January 2016, pp. 23–29.

39. M. N.Bojnordi and E. Ipek, Memristive Boltzmann machine: A hardware accelerator for combinatorial optimization and deep learning, in 2016 IEEE International Symposium on High Performance Computer Architecture (HPCA). IEEE, 2016, pp. 1–13.

40. M. Rafique, B. Lee, and M. Jeon, Hybrid neuromorphic system for automatic speech recognition", *Electronics Letters*, vol. 52, no. 17, pp. 1428–1430, 2016.

41. A. M. Sheri, A. Rafique, W. Pedrycz, and M. Jeon, Contrastive divergence for memristor-based restricted boltzmann machine, *Engineering Applications of Artificial Intelligence*, vol. 37, pp. 336–342, 2015.

42. M. Suri, V. Parmar, A. Kumar, D. Querlioz, and F. Alibart, Neuromorphic hybrid RRAM-CMOS RBM architecture, in 2015 15th Non-Volatile Memory Technology Symposium (NVMTS). IEEE, 2015, pp. 1–6.

43. A. Ascoli, R. Tetzlaff, L. Chua, J. Strachan, and R. Williams, Fading memory effects in a memristor for cellular nanoscale network applications, in 2016 Design, Automation & Test in Europe Conference & Exhibition (DATE). IEEE, 2016, pp. 421–425.

44. F. Corinto, A. Ascoli, Y.-S. Kim, and K.-S.Min, Cellular nonlinear networks with memristor synapses, in A. Adamatsky and L. Chua (eds), *Memristor Networks*. Springer, 2014, pp. 267–291.

45. S. Duan, X. Hu, Z. Dong, L. Wang, and P. Mazumder, Memristorbased cellular nonlinear/neural network: Design, analysis, and applications, IEEE Transactions on Neural Networks and Learning Systems, vol. 26, no. 6, 2015.

46. T. Ibrayev, A. P. James, C. Merkel, and D. Kudithipudi, A design of htm spatial pooler for face recognition using memristor-CMOS hybrid circuits, in 2016 IEEE International Symposium on Circuits and Systems, IEEE, May, 2016.

47. X. Hu, S. Duan, and L. Wang, A novel chaotic neural network using memristive synapse with applications in associative memory, in *Abstract and Applied Analysis, vol. 2012*. Hindawi Publishing Corporation, 2012.

48. T. C. Jackson, R. Shi, A. A. Sharma, J. A. Bain, J. A. Weldon, and L. Pileggi, Implementing delay insensitive oscillatory neural networks using CMOS and emerging technology, *Analog Integrated Circuits and Signal Processing*, vol. 89, no. 3, pp. 619–629, 2016.

49. J. Kang, B. Gao, P. Huang, L. Liu, X. Liu, H. Yu, S. Yu, and H.-S. P. Wong, RRAM based synaptic devices for neuromorphic visual systems, in 2015 IEEE International Conference on Digital Signal Processing (DSP). IEEE, 2015, pp. 1219–1222.

50. X. Zhu, Impact of imprecise programming of memristor on building hardware neural network, in 2011 International Conference on Electrical and Control Engineering (ICECE). IEEE, 2011, pp. 4527–4529.

51. G.Indiveri, R. Legenstein, G. Deligeorgis, T.Prodromakis et al., Integration of nanoscale memristor synapses in neuromorphic computing architectures, *Nanotechnology*, vol. 24, no. 38, p. 384010, 2013.

52. S. Mandal, A. El-Amin, K. Alexander, B. Rajendran, and R. Jha, Novel synaptic memory device for neuromorphic computing, *Scientific Reports*, vol. 4, 2014.

53. H. Wang, H. Li, and R. E. Pino, Memristor-based synapse design and training scheme for neuromorphic computing architecture, in 2012 International Joint Conference on Neural Networks (IJCNN). IEEE, 2012, pp. 1–5.

54. X. Zhu, C. Du, Y. Jeong, and W. D. Lu, Emulation of synaptic metaplasticity in memristors, *Nanoscale*, vol. 9, no. 1, pp. 45–51, 2017.

55. X. Shi, S. Duan, L. Wang, T. Huang, and C. Li, A novel memristive electronic synapsebased hermite chaotic neural network with application in cryptography, *Neurocomputing*, vol. 166, pp. 487–495, 2015.

56. T. C.Jackson, A. A.Sharma, J. A.Bain, J. A.Weldon, and L. Pileggi, An RRAM-based oscillatory neural network, in 2015 IEEE 6th Latin American Symposium on Circuits & Systems (LASCAS). IEEE, 2015, pp. 1–4.

57. T. C.Jackson, A. A.Sharma, J. A.Bain, J. A.Weldon, and L. Pileggi, Oscillatory neural networks based on TMO nano-oscillators and multi-level RRAM cells, *IEEE Journal on Emerging and Selected Topics in Circuits and Systems*, vol. 5, no. 2, pp. 230–241, 2015.

58. A. Sharma, T. Jackson, M. Schulaker, C. Kuo, C. Augustine, J. Bain, H.-S. Wong, S. Mitra, L. Pileggi, and J. Weldon, High performance, integrated 1T1R oxide-based oscillator: Stack engineering for lowpower operation in neural network applications, in 2015 Symposium on VLSI Technology (VLSI Technology). IEEE, 2015, pp. T186–T187.

59. W. Cai and R. Tetzlaff, Synapse as a memristor, in A. Adamatsky and L. Chua (eds), *Memristor Networks*. Springer, 2014, pp. 113–128.

60. T. Chang, S.-H. Jo, K.-H. Kim, P. Sheridan, S. Gaba, and W. Lu, Synaptic behaviors and modeling of a metal oxide memristive device, *Applied Physics A*, vol. 102, no. 4, pp. 857–863, 2011.

61. L. Chua, Memristor, Hodgkin-Huxley, and edge of chaos, *Nanotechnology*, vol. 24, no. 38, p. 383001, 2013.

62. F. Corinto, A. Ascoli, and S.-M. Kang, Memristor-based neural circuits, in 2013 IEEE International Symposium on Circuits and Systems (ISCAS). IEEE, 2013, pp. 417–420.

63. F. Corinto, M. Gilli, A. Ascoli, and R. Tetzlaff, Complex dynamics in neuromorphic memristor circuits, in 2013 European Conference on Circuit Theory and Design (ECCTD). IEEE, 2013, pp. 1–4.

64. S. Gaba, P. Sheridan, J. Zhou, S. Choi, and W. Lu, Stochastic memristive devices for computing and neuromorphic applications, *Nanoscale*, vol. 5, no. 13, pp. 5872–5878, 2013.

65. M. Hu, Y. Wang, W. Wen, Y. Wang, and H. Li, Leveraging stochastic memristor devices in neuromorphic hardware systems, *IEEE Journal on Emerging and Selected Topics in Circuits and Systems*, vol. 12, no. 99, pp. 1–12, 2016.

66. R. Naous, M. AlShedivat, E. Neftci, G. Cauwenberghs, and K. N.Salama, Memristorbased neural networks: Synaptic versus neuronal stochasticity, *AIP Advances*, vol. 6, no. 11, p. 111304, 2016.

67. P. Lorenzi, V. Sucre, G. Romano, R. Rao, and F. Irrera, Memristor based neuromorphic circuit for visual pattern recognition, in 2015 International Conference on Memristive Systems (MEMRISYS). IEEE, 2015, pp. 1–2.

68. A. Wu, Z. Zeng, and J. Xiao, Dynamic evolution evoked by external inputs in memristor-based wavelet neural networks with different memductance functions, *Advances in Difference Equations*, vol. 258, no. 1, pp. 1–14, 2013.

69. A.Aggarwal and B. Hamilton, Training artificial neural networks with memristive synapses: HSPICE-matlab co-simulation, in 2012 11th Symposium on Neural Network Applications in Electrical Engineering (NEUREL). IEEE, 2012, pp. 101–106.

70. A. Ascoli, F. Corinto, M. Gilli, and R. Tetzlaff, Memristor for neuromorphic applications: Models and circuit implementations, in R. Tetzlaff (ed.), *Memristors and Memristive Systems*. Springer, 2014, pp. 379–403.

71. K. D. Cantley, A. Subramaniam, H. J. Stiegler, R. A. Chapman, and E. M. Vogel, Hebbian learning in spiking neural networks with nanocrystalline silicon TFTS and memristive synapses, *IEEE Transactions on Nanotechnology*, vol. 10, no. 5, pp. 1066–1073, 2011.

72. Kurtis D. Cantley, Anand Subramaniam, Harvey J. Stiegler, Richard A. Chapman, and EricM. Vogel, Neural learning circuits utilizing nano-crystalline silicon transistors and memristors, *IEEE Transactions on Neural Networks and Learning Systems*, vol. 23, no. 4, pp. 565–573, 2012.

73. R. Kubendran, Electromagnetic and Laplace domain analysis of memristance and associative learning using memristive synapses modeled in SPICE, in 2012 International Conference on Devices, Circuits and Systems (ICDCS). IEEE, 2012, pp. 622–626.

74. X. Wang, C. Li, T. Huang, and S. Duan, A weakly connected memristive neural network for associative memory, *Neural Processing Letters*, vol. 40, no. 3, pp. 275–288, 2014.

75. S. Wen, Z. Zeng, and T. Huang, Associative learning of integrate-and-fire neurons with memristor-based synapses, *Neural Processing Letters*, vol. 38, no. 1, pp. 69–80, 2013.

76. M. Ziegler and H. Kohlstedt, Mimic synaptic behavior with a single floating gate transistor: A memash synapse, *Journal of Applied Physics*, vol. 114, no. 19, p. 194506, 2013.

77. E. Covi, S. Brivio, M. Fanciulli, and S. Spiga, Synaptic potentiation and depression in al: HfO2-based memristor, *Microelectronic Engineering*, vol. 147, pp. 41–44, 2015.

78. S. Acciarito, A. Cristini, L. DiNunzio, G. M.Khanal, and G. Susi, An a VLSI driving circuit for memristor-based STDP, Ph.D. 2016 12th Conference on Research in Microelectronics and Electronics (PRIME). IEEE, 2016, pp. 1–4.

79. Y. Babacan and F. Kacar, Memristor emulator with spike-timing-dependent-plasticity, *AEU-International Journal of Electronics and Communications*, vol. 73, pp. 16–22, 2017.

80. R. Berdan, E. Vasilaki, A. Khiat, G. Indiveri, A. Serb, and T. Prodromakis, Emulating short-term synaptic dynamics with memristive devices, scientific reports, vol. 6, 2016.

81. J. Bill and R. Legenstein, A compound memristive synapse model for statistical learning through STDP in spiking neural networks, *Frontiers in Neuroscience*, vol. 8, 2014.

82. W. Cai and R. Tetzlaff, Advanced memristive model of synapses with adaptive thresholds, in 2012 13th International Workshop on Cellular Nanoscale Networks and Their Applications (CNNA). IEEE, 2012, pp. 1–6.

83. W. Cai, F. Ellinger, and R. Tetzlaff, Neuronal synapse as a memristor: Modeling pair-and triplet-based STDP rule, *IEEE Transactions on Biomedical Circuits and Systems*, vol. 9, no. 1, pp. 87–95, 2015.

84. K. D.Cantley, A. Subramaniam, H. J.Stiegler, R. Chapman, E. M.Vogel et al., Spike timing-dependent synaptic plasticity using memristors and nano-crystalline silicon tft memories, in 2011 11th IEEE Conference onNanotechnology (IEEE-NANO). IEEE, 2011, pp. 421–425.

85. W. Chan and J. Lohn, Spike timing dependent plasticity with memristive synapse in neuromorphic systems, in 2012 International Joint Conference on Neural Networks (IJCNN). IEEE, 2012, pp. 1–6.

86. L. Chen, C. Li, T. Huang, X. He, H. Li, and Y. Chen, STDP learning rule based onmemristor with STDP property, in 2014 International Joint Conference on Neural Networks (IJCNN). IEEE, 2014, pp. 1–6.

87. E. Covi, S. Brivio, A. Serb, T. Prodromakis, M. Fanciulli, and S. Spiga, Analog memristive synapse in spiking networks implementing unsupervised learning,*Frontiers in Neuroscience*, vol. 10, 2016.

88. J. M.Cruz-Albrecht, T. Derosier, and N. Srinivasa, A scalable neural chip with synapticelectronics using CMOS integrated memristors, *Nanotechnology*, vol. 24, no. 38, p. 384011, 2013.

89. Y. Dai, C. Li, and H. Wang, Expanded HP memristor model and simulation in STDP learning, *Neural Computing and Applications*, vol. 24, no. 1, pp. 51–57, 2014.

90. B. DeSalvo, E. Vianello, O. Thomas, F. Clermidy, O. Bichler, C. Gamrat, and L. Perniola, Emerging resistive memories for low power embedded applications and neuromorphic systems, in 2015 IEEE International Symposium on Circuits and Systems (ISCAS). IEEE, 2015, pp. 3088–3091.

91. N. Du, M. Kiani, C. G. Mayr, T. You, D. Burger, I. Skorupa, O. G.Schmidt, and H. Schmidt, Singlepairing spike-timing dependent plasticity in biFeO3 memristors with a time window of 25 ms to 125 μs, *Frontiers in Neuroscience*, vol. 9, 2015.

92. I. Ebong and P. Mazumder, Memristor based STDP learning network for position detection, in 2010 International Conference on Microelectronics (ICM). IEEE, 2010, pp. 292–295.

93. I. Ebong, D. Deshpande, Y. Yilmaz, and P. Mazumder, Multipurpose neuro-architecture with memristors, in 2011 11th IEEE Conference onNanotechnology (IEEE-NANO). IEEE, 2011, pp. 431–435.

94. I. E. Ebong and P. Mazumder, CMOS and memristor-based neural network design for position detection, *Proceedings of the IEEE*, vol. 100, no. 6, pp. 2050–2060, 2012.

95. E. Gale, B. D. L.Costello, and A. Adamatzky, Observation and characterization of memristor current spikes and their application to neuromorphic computation, in 2012 International Conference of Numerical Analysis and Applied Mathematics (ICNAAM), vol. 1479, no. 1. AIP Publishing, 2012, pp. 1898–1901.

96. W. He, K. Huang, N. Ning, K. Ramanathan, G. Li, Y. Jiang, J. Sze, L. Shi, R. Zhao, and J. Pei, Enabling an integrated rate-temporal learning scheme on memristor, *ScientificReports*, vol. 4, p. 4755, 2014.

97. S. Hu, H. Wu, Y. Liu, T. Chen, Z. Liu, Q. Yu, Y. Yin, and S. Hosaka, Design of an electronic synapse with spike time dependent plasticity based on resistive memory device, *Journal of Applied Physics*, vol. 113, no. 11, p. 114502, 2013.

98. M. Hu, Y. Chen, J. J.Yang, Y. Wang, and H. Li, A memristor-based dynamic synapse for spiking neural networks, *IEEE Transactions on Computer-Aided Design of Integrated Circuits and Systems*, 2016.

99. F. L. M.Huayaney, S. Nease, and E. Chicca, Learning in silicon beyond STDP: A neuromorphic implementation of multi-factor synaptic plasticity with calcium-based dynamics, *IEEE Transactions on Circuits and Systems I: Regular Papers*, vol. 63, no. 12, pp. 2189–2199, 2016.

100. F. L. M. Huayaney and E. Chicca, AVLSI implementation of a calcium-based plasticity learning model, in 2016 IEEE International Symposiumon on Circuits and Systems (ISCAS). IEEE, 2016, pp. 373–376.

101. G. Indiveri and S. Fusi, Spike-based learning in VLSI networks of integrate-and-fire neurons, in 2007 IEEE International Symposium on Circuits and Systems (ISCAS). IEEE, 2007, pp. 3371–3374.

102. S. H. Jo, T. Chang, I. Ebong, B. B.Bhadviya, P. Mazumder, and W. Lu, Nanoscale memristor device as synapse in neuromorphic systems, *Nano Letters*, vol. 10, no. 4, pp. 1297–1301, 2010.

103. Y. Kaneko, Y. Nishitani, M. Ueda, and A. Tsujimura, Neural network based on a three-terminal ferroelectric memristor to enable on-chip pattern recognition, in VLSI Technology 2013 Symposium on (VLSIT). IEEE, 2013, pp. T238–T239.

104. Y. Kaneko, Y. Nishitani, and M. Ueda, Ferroelectric artificial synapses for recognition of a multishaded image, *IEEE Transactions on Electron Devices*, vol. 61, no. 8, pp. 2827–2833, 2014.

105. O. Kavehei, S. Al-Sarawi, K.-R. Cho, N. Iannella, S.-J. Kim, K. Eshraghian, and D. Abbott, Memristor-based synaptic networks and logical operations using in-situ computing, in 2011 Seventh International Conference on Intelligent Sensors, Sensor Networks and Information Processing (ISSNIP). IEEE, 2011, pp. 137–142.

106. G. Lecerf, J. Tomas, and S. Sa.-ghi, Excitatory and inhibitory memristive synapses for spiking neural networks, in 2013 IEEE International Symposium on Circuits and Systems (ISCAS). IEEE, 2013, pp. 1616–1619.

107. Y. Li, L. Xu, Y.-P. Zhong, Y.-X. Zhou, S.-J. Zhong, Y.-Z. Hu, L. O.Chua, and X.-S. Miao, Associative learning with temporal contiguity in a memristive circuit for large-scale neuromorphic networks, *Advanced Electronic Materials*, vol. 1, no. 8, 2015.

108. Q. Li, A. Serb, T. Prodromakis, and H. Xu, A memristor spice model accounting for synaptic activity dependence, *PloS One*, vol. 10, no. 3, p. e0120506, 2015.

109. S. Mandal, B. Long, and R. Jha, Study of synaptic behavior in doped transition metal oxide-based reconfigurable devices, *IEEE Transactions on Electron Devices*, vol. 60, no. 12, pp. 4219–4225, 2013.

110. K. Moon, S. Park, D. Lee, J. Woo, E. Cha, S. Lee, and H. Hwang, Resistive-switching analogue memory device for neuromorphic application, in 2014 IEEE Silicon Nanoelectronics Workshop (SNW). IEEE, 2014, pp. 1–2.

111. H. Mostafa, A. Khiat, A. Serb, C. G.Mayr, G. Indiveri, and T. Prodromakis, Implementation of a spike-based perceptron learning rule using tiO2-x memristors,*Frontiers in Neuroscience*, vol. 9, p. 357, 2015.

112. A. Pantazi, S. Wozniak, T. Tuma, and E. Eleftheriou, All-memristive neuromorphic computing with level-tuned neurons", *Nanotechnology*, vol. 27, no. 35, p. 355205, 2016.

113. M. Payvand, J. Rofeh, A. Sodhi, and L. Theogarajan, A CMOSmemristive selflearning neural network for pattern classification applications, in

2014 IEEE/ACM International Symposium on Nanoscale Architectures (NANOARCH). IEEE, 2014, pp. 92–97.

114. M. Payvand and L. Theogarajan, Exploiting local connectivity of CMOL architecture for highly parallel orientation selective neuromorphic chips, in 2015 IEEE/ACM International Symposium on Nanoscale Architectures (NANOARCH). IEEE, 2015, pp. 187–192.

115. Barranco, On neuromorphic spiking architectures for asynchronous STDP memristive systems, in Proceedings of 2010 IEEE International Symposium onCircuits and Systems (ISCAS). IEEE, 2010, pp. 1659–1662.

116. M. Prezioso, Y. Zhong, D. Gavrilov, F. Merrikh-Bayat, B. Hoskins, G. Adam, K. Likharev, and D. Strukov, Spiking neuromorphic networks with metal-oxide memristors, in 2016 IEEE International Symposium onCircuits and Systems (ISCAS). IEEE, 2016, pp.177–180.

117. K. Seo, I. Kim, S. Jung, M. Jo, S. Park, J. Park, J. Shin, K. P.Biju, J. Kong, K. Lee et al., Analog memory and spike-timing-dependent plasticity characteristics of a nanoscale titanium oxide bilayer resistive switching device, *Nanotechnology*, vol. 22, no. 25, p. 254023, 2011.

118. T. Serrano-Gotarredona and B. Linares-Barranco, Design of adaptive nano/CMOS neural architectures, in 2012 19th IEEE International Conference onElectronics, Circuits and Systems (ICECS). IEEE, 2012, pp. 949–952.

119. T. Serrano-Gotarredona, T. Masquelier, and B. Linares-Barranco, Spike-timingdependent-plasticity with memristors, in A. Adamatsky and L. Chua (eds), *Memristor Networks*. Springer, 2014, pp. 211–247.

120. A. Singha, B. Muralidharan, and B. Rajendran, Analog memristive time dependent learning using discrete nanoscale RRAM devices, in 2014 International Joint Conference on Neural Networks (IJCNN). IEEE, 2014, pp. 2248–2255.

121. G. S.Snider, Spike-timing-dependent learning in memristive nanodevices, in 2008 IEEE International Symposium onNanoscale Architectures (NANOARCH). IEEE, 2008, pp. 85–92.

122. A. Subramaniam, K. D.Cantley, G. Bersuker, D. Gilmer, and E. M.Vogel, Spike-timing-dependent plasticity using biologically realistic action potentials and low-temperature materials, *IEEE Transactions on Nanotechnology*, vol. 12, no. 3, pp. 450–459, 2013.

123. Z. Wang, S. Ambrogio, S. Balatti, and D. Ielmini, A 2-transistor/1-resistor artificial synapse capable of communication and stochastic learning in neuromorphic systems, *Frontiers in Neuroscience*, vol. 8, 2014.

124. T. Werner, E. Vianello, O. Bichler, D. Garbin, D. Cattaert, B. Yvert, B.De Salvo, and L. Perniola, Spiking neural networks based on oxRAM synapses for real-time unsupervised spike sorting, *Frontiers in Neuroscience*, vol. 10, 2016.

125. T. Werner, D. Garbin, E. Vianello, O. Bichler, D. Cattaert, B. Yvert, B. DeSalvo, and L. Perniola, Real-time decoding of brain activity by embedded spiking neural networks using oxRAM synapses, in 2016 IEEE International Symposium on Circuits and Systems (ISCAS). IEEE, 2016, pp. 2318–2321.

126. S. Wozniak, T. Tuma, A. Pantazi, and E. Eleftheriou, Learning spatio-temporal patterns in the presence of input noise using phasechange memristors, in 2016 IEEE International Symposium on Circuits andSystems (ISCAS). IEEE, 2016, pp. 365–368.

127. Y. Zhang, Z. Zeng, and S. Wen, Implementation of memristive neural networks with spike-rate-dependent plasticity synapses, in 2014 International Joint Conference on Neural Networks (IJCNN). IEEE, 2014, pp. 2226–2233.

128. L. Zheng, S. Shin, and S.-M. S. Kang, Memristor-based synapses and neurons for neuromorphic computing, in 2015 IEEE International Symposium on Circuits and Systems (ISCAS). IEEE, 2015, pp. 1150–1153.

129. Q. Wang, Y. Kim, and P. Li, Architectural design exploration for neuromorphic processors with memristive synapses, in 2014 IEEE 14th International Conference on Nanotechnology (IEEE-NANO). IEEE, 2014, pp. 962–966.

130. W. Wang, Z. You, P. Liu, and J. Kuang, *An adaptive neural network a/d converter based on CMOS/memristor hybrid design*, IEICE Electronics Express, 2014.

131. Q. Wang, Y. Kim, and P. Li, Neuromorphic processors with memristive synapses: Synaptic interface and architectural exploration, *ACM Journal on Emerging Technologies in Computing Systems (JETC)*, vol. 12, no. 4, p. 35, 2016.

132. R. Hasan and T. M. Taha, Enabling back propagation training of memristor crossbar neuromorphic processors, in 2014 International Joint Conference on Neural Networks (IJCNN). IEEE, 2014, pp. 21–28.

133. M. Prezioso, I. Kataeva, F. Merrikh-Bayat, B. Hoskins, G. Adam, T. Sota, K. Likharev, and D. Strukov, Modeling and implementation of firing-rate neuromorphic-network classifiers with bilayer Pt/AL2O3/TiO2-x/Pt memristors, in 2015 IEEE International Electron Devices Meeting (IEDM). IEEE, 2015, pp. 17–4.

134. E. Zamanidoost, M. Klachko, D. Strukov, and I. Kataeva, Low area overhead in-situ training approach for memristor-based classifier, in 2015 IEEE/ACM International Symposium on Nanoscale Architectures (NANOARCH). IEEE, 2015, pp. 139–142.

135. B.Li, Y. Wang, Y. Wang, Y. Chen, and H. Yang, Training itself: Mixed-signal training acceleration for memristor-based neural network, in 2014

19th Asia and South Pacific Design Automation Conference (ASP-DAC). IEEE, 2014, pp. 361–366.

136. M. V. Nair and P. Dudek, Gradient-descent-based learning in memristive crossbar arrays, in 2015 International Joint Conference on Neural Networks (IJCNN). IEEE, 2015, pp. 1–7.

137. C. H. Bennett, S. La Barbera, A. F. Vincent, J.-O. Klein, F. Alibart, and D. Querlioz, Exploiting the short-term to long-term plasticity transition in memristive nanodevice learning architectures, in 2016 International Joint Conference on Neural Networks (IJCNN). IEEE, 2016, pp. 947–954.

138. D. Chabi, W. Zhao, D. Querlioz, and J.-O. Klein, Robust neural logic block (NLB) based on memristor crossbar array, in 2011 IEEE/ACM International Symposium on Nanoscale Architectures (NANOARCH). IEEE, 2011, pp. 137–143.

139. D. Chabi, D. Querlioz, W. Zhao, and J.-O. Klein, Robust learning approach for neuro-inspired nanoscale crossbar architecture, *ACM Journal on Emerging Technologies in Computing Systems (JETC)*, vol. 10, no. 1, p. 5, 2014.

140. D. Chabi, Z. Wang, C. Bennett, J.-O. Klein, and W. Zhao, Ultra high density memristor neural crossbar for on-chip supervised learning, IEEE Transactions on Nanotechnology, vol. 14, no. 6, pp. 954–962, 2015.

141. D. Chabi, W. Zhao, D. Querlioz, and J.-O. Klein, On-chip universal supervised learning methods for neuro-inspired block of memristive nanodevices, *ACM Journal on Emerging Technologies in Computing Systems (JETC)*, vol. 11, no. 4, p. 34, 2015.

142. M. Hu, H. Li, Y. Chen, Q. Wu, G. S. Rose, and R. W. Linderman, Memristor crossbar-based neuromorphic computing system: A case study, *IEEE Transactions onNeural Networks and Learning Systems*, vol. 25, no. 10, pp. 1864–1878, 2014.

143. Q. Qiu, Z. Li, K. Ahmed, H. Li, and M. Hu, Neuromorphic acceleration for context aware text image recognition, in 2014 IEEE Workshop on Signal Processing Systems (SiPS). IEEE, 2014, pp. 1–6.

144. Q. Qiu, Z. Li, K. Ahmed, W. Liu, S. F. Habib, H. H. Li, and M. Hu, A neuromorphic architecture for context aware text image recognition, *Journal of Signal Processing Systems*, vol. 84, no. 355, pp. 1–15, 2015.

145. T. M. Taha, R. Hasan, and C. Yakopcic, Memristor crossbar based multicore neuromorphic processors, in 2014 27th IEEE International System-on-Chip Conference (SOCC). IEEE, 2014, pp. 383–389.

146. F. Merrikh-Bayat and S. B. Shouraki, The neuro-fuzzy computing system with the capacity of implementation on a memristor crossbar and optimization-free hardware training, *IEEE Transactions on Fuzzy Systems*, vol. 22, no. 5, pp. 1272–1287, 2014.

147. M. R. Azghadi, B. Linares-Barranco, D. Abbott, and P. H. Leong, A hybrid CMOSmemristor neuromorphic synapse, *IEEE Transactions on Biomedical Circuits and Systems*, 2016.

148. P.-Y. Chen, L. Gao, and S. Yu, Design of resistive synaptic array for implementing on-chip sparse learning, IEEE Transactions on MultiScale Computing Systems, vol. 2, no. 4, pp. 257–264, 2016.

149. M. Chu, B. Kim, S. Park, H. Hwang, M. Jeon, B. H.Lee, and B.-G. Lee, Neuromorphic hardware system for visual pattern recognition with memristor array and CMOS neuron, *IEEE Transactions on Industrial Electronics*, vol. 62, no. 4, pp. 2410–2419, 2015.

150. Y. Kim, Y. Zhang, and P. Li, A digital neuromorphic VLSI architecture with memristor crossbar synaptic array for machine learning, in 2012 IEEE International SOC Conference (SOCC). IEEE, 2012, pp. 328–333.

151. Y. Kim, Y. Zhang, and P. Li, A reconfigurable digital neuromorphic processor with memristive synaptic crossbar for cognitive computing, *ACM Journal on Emerging Technologies in Computing Systems (JETC)*, vol. 11, no. 4, p. 38, 2015.

152. G. Li, L. Deng, D. Wang, W. Wang, F. Zeng, Z. Zhang, H. Li, S. Song, J. Pei, and L. Shi, Hierarchical chunking of sequential memory on neuromorphic architecture with reduced synaptic plasticity, *Frontiers in Computational Neuroscience*, vol. 10, 2016.

153. B. Linares-Barranco and T. Serrano-Gotarredona, Exploiting memristance in adaptive asynchronous spiking neuromorphic nanotechnology systems, in 2009 IEEE-NANO 9th IEEE Conference on Nanotechnology. IEEE, 2009, pp. 601–604.

154. H. Mostafa, C. Mayr, and G. Indiveri, Beyond spike-timing dependent plasticity in memristor crossbar arrays, 2016 IEEE International Symposium on Circuits and Systems (ISCAS). IEEE, 2016, pp. 926–929.

155. R. Naous, M. Al-Shedivat, E. Neftci, G. Cauwenberghs, and K. N.Salama, Stochastic synaptic plasticity with memristor crossbar arrays, in 2016 IEEE International Symposium on Circuits and Systems (ISCAS). IEEE, 2016, pp. 2078–2081.

156. S. Park, H. Kim, M. Choo, J. Noh, A. Sheri, S. Jung, K. Seo, J. Park, S. Kim, W. Lee et al., RRAM-based synapse for neuromorphic system with pattern recognition function, in Electron Devices Meeting (IEDM), 2012, pp. 10–2.

157. S. Park, A. Sheri, J. Kim, J. Noh, J. Jang, M. Jeon, B. Lee, B. Lee, B. Lee, and H.-j. Hwang, Neuromorphic speech systems using advanced reram-based synapse, *IEDM Tech Dig*, vol. 25, pp. 1–25, 2013.

158. M. Prezioso, F. Merrikh-Bayat, B. Chakrabarti, and D. Strukov, RRAM-based hardware implementations of artificial neural networks: progress

update and challenges ahead, in SPIE OPTO. *International Society for Optics and Photonics*, 2016, pp. 918–974.

159. M. Prezioso, F. M.Bayat, B. Hoskins, K. Likharev, and D. Strukov, Self-adaptive spike-time-dependent plasticity of metal-oxide memristors, *Scientific Reports*, vol. 6, 2016.

160. M. Prezioso, Experimental analog implementation of neural networks on integrated metal-oxide memristive crossbar arrays, in 2016 IEEE 16th International Conference onNanotechnology (IEEE-NANO). IEEE, 2016, pp. 276–279.

161. Q. Ren, Q. Long, Z. Zhang, and J. Zhao, Information transfer characteristic in memristic neuromorphic network, in C. Guo, Z.-G. Hou, and Z. Zeng, Advances in Neural Networks-ISNN 2013. Springer, 2013, pp. 1–8.

162. A. M.Sheri, H. Hwang, M. Jeon, and B.-g. Lee, Neuromorphic character recognition system with two PCMO memristors as a synapse, *IEEE Transactions on Industrial Electronics*, vol. 61, no. 6, pp. 2933–2941, 2014.

163. Z. Wang, W. Zhao, W. Kang, Y. Zhang, J.-O. Klein, and C. Chappert, Ferroelectric tunnel memristor-based neuromorphic network with 1T1R crossbar architecture, in 2014 International Joint Conference on Neural Networks (IJCNN). IEEE, 2014, pp. 29–34.

164. X. Wu, V. Saxena, and K. Zhu, Homogeneous spiking neuromorphic system for real-world pattern recognition, IEEE Journal on Emerging and Selected Topics in Circuits and Systems, vol. 5, pp. 254–266, 2015.

165. X. Wu, V. Saxena, and K. Zhu, A CMOS spiking neuron for dense memristor-synapse connectivity for brain inspired computing, in 2015 International Joint Conference on Neural Networks (IJCNN). IEEE, 2015, pp. 1–6.

166. X. Wu, V. Saxena, K. Zhu, and S. Balagopal, A CMOS spiking neuron for brain-inspired neural networks with resistive synapses and in situ learning, *IEEE Transactions on Circuits and Systems II: Express Briefs*, vol. 62, no. 11, pp. 1088–1092, 2015.

167. X. Yang, W. Chen, and F. Z.Wang, A supervised spiking time dependant plasticity network based on memristors, in 2013 IEEE 14th International Symposium on Computational Intelligence and Informatics (CINTI). IEEE, 2013, pp. 447–451.

168. S. N.Truong, K. VanPham, W. Yang, K.-S. Min, Y. Abbas, and C. J.Kang, Live demonstration: Memristor synaptic array with FPGA-implemented neurons for neuromorphic pattern recognition, in 2016 IEEE Asia Pacific Conference on Circuits and Systems (APCCAS). IEEE, 2016, pp. 742–743.

169. S. N. Truong, K. Van Pham, W. Yang, J. Song, H.-S. Mo, and K.-S. Min, FPGA-based training and recalling system for memristor synapses, in 2016 International Conference on Electronics, Information, and Communications (ICEIC). IEEE, 2016, pp. 1–4.

170. G. C.Adam, B. D.Hoskins, M. Prezioso, F. Merrikh-Bayat, B. Chakrabarti, and D. B.Strukov, 3-D memristor crossbars for analog and neuromorphic computing applications, *IEEE Transactions on Electron Devices*, vol. 64, no. 1, pp. 312–318, 2017.

171. S. Agarwal, T.-T. Quach, O. Parekh, A. H. Hsia, E. P. DeBenedictis, C. D. James, M. J. Marinella, and J. B. Aimone, Energy scaling advantages of resistive memory crossbar based computation and its application to sparse coding, *Frontiers in Neuroscience*, vol. 9, 2015.

172. M. Bavandpour, H. Soleimani, B. Linares-Barranco, D. Abbott, and L. O.Chua, Generalized reconfigurable memristive dynamical system (MDS) for neuromorphic applications, *Frontiers in Neuroscience*, vol. 9, 2015.

173. D. Chabi, Z. Wang, W. Zhao, and J.-O. Klein, On-chip supervised learning rule for ultra-high density neural crossbar using memristor for synapse and neuron, in Proceedingsof the 2014 IEEE/ACM International Symposium on Nanoscale Architectures. ACM, 2014, pp. 7–12.

174. Q. Chen, Q. Qiu, H. Li, and Q. Wu, A neuromorphic architecture for anomaly detection in autonomous large-area traffic monitoring, in 2013 IEEE/ACM International Conference on Computer-Aided Design (ICCAD). IEEE, 2013, pp. 202–205.

175. P.-Y. Chen and S. Yu, Partition SRAM and RRAM based synaptic arrays for neuro-inspired computing, in 2016 IEEE International Symposium on Circuits and Systems (ISCAS). IEEE, 2016, pp. 2310–2313.

176. P. Chi, S. Li, C. Xu, T. Zhang, J. Zhao, Y. Liu, Y. Wang, and Y. Xie, Prime: A novel processing-in-memory architecture for neural network computation in reram-based mainmemory, in Proceedings of the 43rd International Symposium on Computer Architecture. IEEE Press, 2016, pp. 27–39.

177. H. Choi, H. Jung, J. Lee, J. Yoon, J. Park, D.J. Seong, W. Lee, M. Hasan, G.-Y. Jung, and H. Hwang, An electrically modifiable synapse array of resistive switching memory, *Nanotechnology*, vol. 20, no. 34, p. 345201, 2009.

178. R. Hasan, C. Yakopcic, and T. M.Taha, Ex-situ training of dense memristor crossbar for neuromorphic applications, in 2015 IEEE/ACM International Symposium on Nanoscale Architectures (NANOARCH). IEEE, 2015, pp. 75–81.

179. M. Hu, H. Li, Q. Wu, and G. S. Rose, Hardware realization of BSB recall function using memristor crossbar arrays, in Proceedings of the 49th Annual Design Automation Conference. ACM, 2012, pp. 498–503.

180. M. Hu, J. P. Strachan, Z. Li, E. M. Grafals, N. Davila, C. Graves, S. Lam, N. Ge, R.S. Williams, and J. Yang, Dot-product engine for neuromorphic computing: Programming 1T1M crossbar to accelerate matrix-vector multiplication, in Proceedings of DAC, vol. 53, 2016.

181. H. Jiang, W. Zhu, F. Luo, K. Bai, C. Liu, X. Zhang, J. J.Yang, Q. Xia, Y. Chen, and Q. Wu, Cyclical sensing integrate-and-fire circuit for memristor array

based neuromorphic computing, in 2016 IEEE International Symposium on Circuits and Systems (ISCAS). IEEE, 2016, pp. 930–933.

182. J. Kataeva, F. Merrikh-Bayat, E. Zamanidoost, and D. Strukov, Efficient training algorithms for neural networks based on memristive crossbar circuits, in 2015 International Joint Conference on Neural Networks (IJCNN). IEEE, 2015, pp. 1–8.

183. K.-H.Kim, S. Gaba, D. Wheeler, J. M.Cruz-Albrecht, T. Hussain, N. Srinivasa, and W. Lu, A functional hybrid memristor crossbararray/CMOS system for data storage and neuromorphic applications, *Nano Letters*, vol. 12, no. 1, pp. 389–395, 2011.

184. H. Li, X. Liu, M. Mao, Y. Chen, Q. Wu, and M. Bamell, Neuromorphic hardware acceleration enabled by emerging technologies, in 2014 14th International Symposium on Integrated Circuits (ISIC). IEEE, 2014, pp. 124–127.

185. Z. Li, C. Liu, Y. Wang, B. Yan, C. Yang, J. Yang, and H. Li, An overview on memristor crossabr based neuromorphic circuit and architecture, in 2015 IFIP/IEEE International Conference on Very Large Scale Integration (VLSI-SoC). IEEE, 2015, pp. 52–56.

186. H. Li, B. Liu, X. Liu, M. Mao, Y. Chen, Q. Wu, and Q. Qiu, The applications of memristor devices in next-generation cortical processor designs, in 2015 IEEE International Symposium on Circuits and Systems (ISCAS). IEEE, 2015, pp. 17–20.

187. H. H.Li, C. Liu, B. Yan, C. Yang, L. Song, Z. Li, Y. Chen, W. Zhu, Q. Wu, and H. Jiang, Spiking-based matrix computation by leveraging memristor crossbar array, in 2015 IEEE Symposium on Computational Intelligence for Security and Defense Applications (CISDA). IEEE, 2015, pp. 1–4.

188. B. Liu, M. Hu, H. Li, Z.-H. Mao, Y. Chen, T. Huang, and W. Zhang, Digital-assisted noise-eliminating training for memristor crossbarbased analog neuromorphic computing engine, in Design Automation Conference (DAC), 2013 50th ACM/EDAC/IEEE. IEEE, 2013, pp. 1–6.

189. X. Liu, M. Mao, H. Li, Y. Chen, H. Jiang, J. J.Yang, Q. Wu, and M. Barnell, A heterogeneous computing system with memristor based neuromorphic accelerators, in 2014 IEEE High Performance Extreme Computing Conference (HPEC). IEEE, 2014, pp. 1-6.

190. B. Liu, H. Li, Y. Chen, X. Li, T. Huang, Q. Wu, and M. Barnell, Reduction and IR-drop compensations techniques for reliable neuromorphic computing systems, in 2014 IEEE/ACM International Conference on Computer-Aided Design (ICCAD). IEEE, 2014, pp. 63–70.

191. X. Liu, M. Mao, B. Liu, H. Li, Y. Chen, B. Li, Y. Wang, H. Jiang, M. Barnell, Q. Wu et al., Reno: A high-efficient reconfigurable neuromorphic computing accelerator design, in Design Automation Conference (DAC), 2015 52ndACM/EDAC/IEEE. IEEE, 2015, pp.1–6.

192. C. Liu, B. Yan, C. Yang, L. Song, Z. Li, B. Liu, Y. Chen, H. Li, Q. Wu, and H. Jiang, A spiking neuromorphic design with resistive crossbar, in Proceedings of the 52nd Annual Design Automation Conference. ACM, 2015, p. 14.

193. B. Liu, W. Wen, Y. Chen, X. Li, C.-R. Wu, and T.-Y. Ho, Eda challenges for memristor-crossbar based neuromorphic computing, in Proceedings of the 25th edition on Great Lakes Symposium on VLSI. ACM, 2015, pp. 185–188.

194. X. Liu, M. Mao, B. Liu, B. Li, Y. Wang, H. Jiang, M. Barnell, Q. Wu, J. Yang, H. Li et al., Harmonica: A framework of heterogeneous computing systems with memristor-based neuromorphic computing accelerators, *IEEE Transactions on Circuits and Systems I: Regular Papers*, vol. 63, no. 5, pp. 617–628, 2016.

195. C. Liu, Y. Chen, and H. Li, Neural processor design enabled by memristor technology, in IEEE International Conference on Rebooting Computing (ICRC). IEEE, 2016, pp. 1–4.

196. Y. Long, E. M.Jung, J. Kung, and S. Mukhopadhyay, ReRAM crossbar based recurrent neural network for human activity detection, in 2016 International Joint Conference on Neural Networks (IJCNN). IEEE, 2016, pp. 939–946.

197. C. Merkel and D. Kudithipudi, Unsupervised learning in neuromemristive systems, in 2015 National Aerospace and Electronics Conference (NAECON). IEEE, 2015, pp. 336–338.

198. S. Park, J. Noh, M. Choo, A. M. Sheri, M. Chang, Y.-B. Kim, C. J. Kim, M. Jeon, B.-G. Lee, B. H. Lee et al., Nanoscale RRAM based synaptic electronics: Toward a neuromorphic computing device, *Nanotechnology*, vol. 24, no. 38, p. 384009, 2013.

199. S. Park, M. Chu, J. Kim, J. Noh, M. Jeon, B. H.Lee, H. Hwang, B. Lee, and B.-g. Lee, Electronic system with memristive synapses for pattern recognition, *Scientific Reports*, vol. 5, 2015.

200. M. Prezioso, F. Merrikh-Bayat, B. Hoskins, G. Adam, K. K. Likharev, and D. B. Strukov, Training and operation of an integrated neuromorphic network based on metaloxide memristors, *Nature*, vol. 521, no. 7550, pp. 61–64, 2015.

201. F. Rothganger, B. R. Evans, J. B. Aimone, and E. P. DeBenedictis, Training neural hardware with noisy components, in 2015 International Joint Conference on Neural Networks (IJCNN). IEEE, 2015, pp. 1–8.

202. A. Sengupta and K. Roy, Spin-transfer torque magnetic neuron for low power neuromorphic computing, in 2015 International Joint Conference on Neural Networks (IJCNN). IEEE, 2015, pp. 1–7.

203. A. Shafiee, A. Nag, N. Muralimanohar, R. Balasubramanian, J. Strachan, M. Hu, R. S. Williams, and V. Srikumar, *Isaac: A convolutional neural network accelerator with in-situanalog arithmetic in crossbars*. ISCA, 2016.

204. J. A.Starzyk et al., Comparison of two memristor based neural network learning schemes for crossbar architecture, in I. Rojas, G. Joya, and J. Gabestany, *Advances in Computational Intelligence*. Springer, 2013, pp. 492–499.
205. J. Starzyk et al., Memristor crossbar architecture for synchronous neural networks, *IEEE Transactions onCircuits and Systems I: Regular Papers*, vol. 61, no. 8, pp. 2390–2401, 2014.
206. M. S. Tarkov, Mapping neural network computations onto memristor crossbar, in 2015 International Siberian Conference on Control and Communications (SIBCON). IEEE, 2015, pp. 1–4.
207. M. Tarkov, Mapping weight matrix of a neural network's layer onto memristor crossbar, *Optical Memory and Neural Networks*, vol. 24, no. 2, pp. 109–115, 2015.
208. S. N.Truong, S.-J. Ham, and K.-S. Min, Neuromorphic crossbar circuit with nanoscale filamentary-switching binary memristors for speech recognition, *Nanoscale Research Letters*, vol. 9, no. 1, pp. 1–9, 2014.
209. S. N. Truong and K.-S. Min, New memristor-based crossbar array architecture with 50-matrix-vector multiplication of analog neuromorphic computing, *Journal of Semiconductor Technology and Science*, vol. 14, no. 3, pp. 356–363, 2014.
210. S. N. Truong, K. Van Pham, W. Yang, and K.-S. Min, Sequential memristor crossbar for neuromorphic pattern recognition, *IEEE Transactions on Nanotechnology*, vol. 15, no. 6, pp. 922–930, 2016.
211. S. Wang, W. Wang, C. Yakopcic, E. Shin, G. Subramanyam, and T. Taha, Reconfigurable neuromorphic crossbars based on titanium oxide memristors, *Electronics Letters*, vol. 52, no. 20, pp. 1673–1675, 2016.
212. D. Wheeler, K. Kim, S. Gaba, E. Wang, S. Kim, I. Valles, J. Li, Y. Royter, J. Cruz-Albrecht, T. Hussain et al., CMOS-integrated memristors for neuromorphic architectures, in *Semiconductor Device Research Symposium (ISDRS)*, 2011 International. IEEE, 2011, pp. 1–2.
213. D. Wheeler, I. Alvarado-Rodriguez, K. Elliott, J. Kally, J. Hermiz, H. Hunt, T. Hussain, and N. Srinivasa, Fabrication and characterization of tungsten-oxide-based memristors for neuromorphic circuits, in 2014 14th International Workshop on Cellular Nanoscale Networks and their Applications (CNNA). IEEE, 2014, pp. 1–2.
214. L.Xia, T. Tang, W. Huangfu, M. Cheng, X. Yin, B. Li, Y. Wang, and H. Yang, Switched by input: Power efficient structure for RRAM-based convolutional neural network, in Proceedings of the 53rd Annual Design Automation Conference. ACM, 2016, p. 125.
215. G. Xie, G. Liu, and S. Zhang, Expression of emotion using a system combined artificial neural network and memristor-based crossbar

array, in 2016 35th Chinese Control Conference (CCC). IEEE, 2016, pp. 9837–9841.

216. L. Xu, C. Li, and L. Chen, Analog memristor based neuromorphic crossbar circuit for image recognition, in 2015 Sixth International Conference on Intelligent Control and Information Processing (ICICIP). IEEE, 2015, pp. 155–160.

217. C. Yakopcic and T. M. Taha, Energy efficient perceptron pattern recognition using segmented memristor crossbar arrays, in 2013 International Joint Conference on Neural Networks (IJCNN). IEEE, 2013, pp. 1–8.

218. C. Yakopcic, R. Hasan, and T. M. Taha, Tolerance to defective memristors in a neuromorphic learning circuit, in 2014 IEEE National Aerospace and Electronics Conference, NAECON. IEEE, 2014, pp. 243–249

219. C. Yakopcic, R. Hasan, T. M. Taha, M. R. McLean, and D. Palmer, Efficacy of memristive crossbars for neuromorphic processors, in 2014 International Joint Conference on Neural Networks (IJCNN). IEEE, 2014, pp. 15–20.

220. C. Yakopcic, R. Hasan, T. M. Taha, M. McLean, and D. Palmer, Memristor-based neuron circuit and method for applying learning algorithm in spice? *Electronics Letters*, vol. 50, no. 7, pp. 492–494, 2014.

221. C. Yakopcic, T. Taha, and M. McLean, Method for ex-situ training in memristor-based neuromorphic circuit using robust weight programming method, *Electronics Letters*, vol. 51, no. 12, pp. 899–900, 2015.

222. C. Yakopcic, R. Hasan, and T. M.Taha, Memristor based neuromorphic circuit for ex-situ training of multi-layer neural network algorithms, in 2015 International Joint Conference on Neural Networks (IJCNN). IEEE, 2015, pp. 1–7.

223. C. Yakopcic, T. M.Taha, G. Subramanyam, and R. E.Pino, Impact of memristor switching noise in a neuromorphic crossbar, in 2015 National Aerospace and Electronics Conference (NAECON). IEEE, 2015, pp. 320–326.

224. C. Yakopcic, R. Hasan, T. M.Taha, and D. Palmer, Spice analysis of dense memristor crossbars for low power neuromorphic processor designs, in 2015 National Aerospace and Electronics Conference (NAECON). IEEE, 2015, pp. 305–311.

225. C. Yakopcic and T. M.Taha, Ex-situ programming in a neuromorphic memristor based crossbar circuit, in 2015 National Aerospace and Electronics Conference (NAECON). IEEE, 2015, pp. 300–304.

226. C. Yakopcic, M. Z.Alom, and T. M.Taha, Memristor crossbar deep network implementation based on a convolutional neural network, in 2016 International Joint Conference on Neural Networks (IJCNN). IEEE, 2016, pp. 963–970.

227. B. Yan, A. M.Mahmoud, J. J.Yang, Q. Wu, Y. Chen, and H. H.Li, A neuromorphic asic design using one-selector-one-memristor crossbar, in 2016 IEEE International Symposium on Circuits and Systems (ISCAS). IEEE, 2016, pp. 1390–1393.

228. P. Yao, H. Wu, B. Gao, G. Zhang, and H. Qian, The effect of variation on neuromorphic network based on 1T1R memristor array, in 2015 15th Non-Volatile Memory Technology Symposium (NVMTS). IEEE, 2015, pp. 1–3.

229. K. Yogendra, D. Fan, and K. Roy, Coupled spin torque nano oscillators for low power neural computation,*IEEE Transactions on Magnetics,* vol. 51, no. 10, pp. 1–9, 2015

230. S. Yu, B. Gao, Z. Fang, H. Yu, J. Kang, and H.-S. Wong, A neuromorphic visual system using RRAM synaptic devices with SUBPJ energy and tolerance to variability: Experimental characterization and large-scale modeling, in 2012 IEEE International Electron Devices Meeting (IEDM). IEEE, 2012, pp. 10–4.

231. S. Yu, P.-Y. Chen, Y. Cao, L. Xia, Y. Wang, and H. Wu, Scaling-up resistive synaptic arrays for neuro-inspired architecture: Challenges and prospect, in 2015 IEEE International Electron Devices Meeting (IEDM). IEEE, 2015, pp. 17–3.

232. L. Chen, C. Li, T. Huang, Y. Chen, S. Wen, and J. Qi, A synapse memristor model with forgetting effect, *Physics Letters A*, vol. 377, no. 45, pp. 3260–3265, 2013.

233. P. Zhang, C.Li, T. Huang, L. Chen, and Y.Chen, Forgetting memristor basedneuromorphic system for pattern training and recognition, *Neurocomputing*, vol. 222, pp. 47–53, 2017.

234. C. Merkel, D. Kudithipudi, and N. Sereni, Periodic activation functions in memristorbased analog neural networks, in 2013 International Joint Conference on Neural Networks (IJCNN). IEEE, 2013, pp. 1–7.

235. D. Soudry, D. Di Castro, A. Gal, A. Kolodny, and S. Kvatinsky, Memristor-based multilayer neural networks with online gradient descent training IEEE transactions on neural networks and learning systems, vol. 26, no. 10, pp. 2408–2421, 2015.

236. E. Rosenthal, S. Greshnikov, D. Soudry, and S. Kvatinsky, A fully analog memristorbased neural network with online gradient training, in 2016 IEEE International Symposium on Circuits and Systems (ISCAS). IEEE, 2016, pp. 1394–1397.

237. Y. Liu, H. Huang, and T. Huang, Associate learning law in a memristive neural network, in 2013 Sixth International Conference on Advanced Computational Intelligence (ICACI). IEEE, 2013, pp. 212–217.

238. G. Howard, E. Gale, L. Bull, B. de Lacy Costello, and A. Adamatzky, Towards evolving spiking networks with memristive synapses, in 2011 IEEE Symposium on Artificial Life (ALIFE). IEEE, 2011, pp. 14–21.

239. G. Howard, E. Gale, L. Bull, B. de Lacy Costello, and A. Adamatzky, Evolution of plastic learning in spiking networks via memristive connections, *IEEE Transactions on Evolutionary Computation*, vol. 16, no. 5, pp. 711–729, 2012.

240. G.Howard, L. Bull, B. de Lacy Costello, E. Gale, and A. Adamatzky, Evolving spiking networks with variable resistive memories,*Evolutionary Computation*, vol. 22, no. 1, pp.79–103, 2014.

241. G. D. Howard, L. Bull, B. D. L. Costello, E. Gale, and A. Adamatzky, Evolving memristive neural networks, in A. Adamatsky and L. Chua (eds), *Memristor Networks*. Springer, 2014, pp. 293–322.

242. D. Howard, L. Bull, and B. de Lacy Costello, Evolving unipolar memristor spiking neural networks, in S. K. Chalup, A. D. Blair and M. Randall (eds), *Artificial Life and Computational Intelligence*. Springer, 2015, pp. 258–272.

243. M. Soltiz, C. Merkel, D. Kudithipudi, and G. S. Rose, RRAM-based adaptive neural logic block for implementing non-linearly separable functions in a single layer, in 2012 IEEE/ACM International Symposium on Nanoscale Architectures (NANOARCH). IEEE, 2012, pp. 218–225.

244. P.-Y. Chen, B. Lin, I. Wang, T.-H. Hou, J. Ye, S. Vrudhula, J.-s. Seo, Y. Cao, S. Yu et al., Mitigating effects of non-ideal synaptic device characteristics for on-chip learning, in 2015 IEEE/ACM International Conference on Computer-Aided Design (ICCAD). IEEE, 2015, pp. 194–199.

245. I.Ebong and P. Mazumder, Memristor based STDP learning network for position detection, in 2010 International Conference on Microelectronics (ICM). IEEE, 2010, pp. 292–295.

246. I. E.Ebong and P. Mazumder, CMOS and memristor-based neural network design for position detection, *Proceedings of the IEEE*, vol. 100, no. 6, pp. 2050–2060, 2012.

247. D. Querlioz, W. Zhao, P. Dollfus, J. Klein, O. Bichler, and C. Gamrat, Bioinspired networks with nanoscale memristive devices that combine the unsupervised and supervised learning approaches, in 2012 IEEE/ACM International Symposium on Nanoscale Architectures (NANOARCH). IEEE, 2012, pp. 203–210.

248. M. Shahsavari, P. Falez, and P. Boulet, Combining a volatile and nonvolatile memristor in artificial synapse to improve learning in spiking neural networks, in 2016 IEEE/ACM International Symposium on Nanoscale Architectures (NANOARCH). IEEE, 2016, pp. 67–72.

249. Z. Chen, B. Gao, Z. Zhou, P. Huang, H. Li, W. Ma, D. Zhu, L. Liu, X. Liu, J. Kang et al., Optimized learning scheme for grayscale image recognition in a RRAM based analog neuromorphic system, in 2015 IEEE International Electron Devices Meeting (IEDM). IEEE, 2015, pp. 17–7.

250. S. Danilin and S. Shchanikov, Neural network control over operation accuracy of memristor-based hardware, in 2015 International Conference on Mechanical Engineering, Automation and Control Systems (MEACS). IEEE, 2015, pp. 1–5.

251. S. Danilin, S. Shchanikov, and A. Galushkin, The research of memristor-based neural network components operation accuracy in control and communication systems, in 2015 International Siberian Conference on Control and Communications (SIBCON). IEEE, 2015, pp. 1–6.

252. L. Deng, G. Li, N. Deng, D. Wang, Z. Zhang, W. He, H. Li, J. Pei, and L. Shi, Complex learning in bio-plausible memristive networks, *Scientific Reports*, vol. 5, 2015.

253. D. Querlioz, P. Dollfus, O. Bichler, and C. Gamrat, Learning with memristive devices: How should we model their behavior? in 2011 IEEE/ACM International Symposium on Nanoscale Architectures (NANOARCH). IEEE, 2011, pp. 150–156.

254. G. S.Rose, R. Pino, and Q. Wu, A low-power memristive neuromorphic circuit utilizing a global/local training mechanism, in 2011 international joint conference on Neural networks (IJCNN). IEEE, 2011, pp. 2080–2086.

255. M. Soltiz, D. Kudithipudi, C. Merkel, G. S.Rose, and R. E.Pino, Memristor-based neural logic blocks for nonlinearly separable functions, , *IEEE Transactions on Computers*, vol. 62, no. 8, pp. 1597–1606, 2013.

256. W. Woods, J. R.Burger, and C. Teuscher, On the inuence of synaptic weight states in a locally competitive algorithm for memristive hardware, in 2014 IEEE/ACM International Symposium on Nanoscale Architectures (NANOARCH). IEEE, 2014, pp. 19–24.

257. W. Woods, J. Burger, and C. Teuscher, Synaptic weight states in a locally competitive algorithm for neuromorphic memristive hardware, *IEEE Transactions on Nanotechnology*, vol. 14, pp. 945–953, 2015.

258. C.-R. Wu, W. Wen, T.-Y. Ho, and Y. Chen, Thermal optimization for memristor-based hybrid neuromorphic computing systems, in 2016 21st Asia and South Pacific Design Automation Conference (ASPDAC). IEEE, 2016, pp. 274–279.

259. A. Pantazi, S. Wozniak, T. Tuma, and E. Eleftheriou, All-memristive neuromorphic computing with level-tuned neurons, *Nanotechnology*, vol. 27, no. 35, p. 355205, 2016.

260. M. Al-Shedivat, R. Naous, E. Neftci, G. Cauwenberghs, and K. N.Salama, Inherently stochastic spiking neurons for probabilistic neural computation, in 2015 7th International IEEE/EMBS Conference on Neural Engineering (NER). IEEE, 2015, pp. 356–359.

261. M. Al-Shedivat, R. Naous, G. Cauwenberghs, and K. N.Salama, Memristors empower spiking neurons with stochasticity, *IEEE Journal on Emerging and Selected Topics in Circuits and Systems*, vol. 5, no. 2, 2015.

262. Y. Babacan, F. Kaçar, and K. Gurkan, A spiking and bursting neuron circuit based on memristor, *Neurocomputing*, 2016.

263. V. Demin, V. Erokhin, A. Emelyanov, S. Battistoni, G. Baldi, S. Iannotta, P. Kashkarov, and M. Kovalchuk, Hardware elementary perceptron based on polyaniline memristivedevices, *Organic Electronics*, vol. 25, pp. 16–20, 2015.

264. A. Mehonic and A. J.Kenyon, Emulating the electrical activity of the neuron using a silicon oxide RRAM cell, *Frontiers in Neuroscience*, vol. 10, 2016.

265. J. Shamsi, A. Amirsoleimani, S. Mirzakuchaki, and M. Ahmadi, Modular neuron comprises of memristor-based synapse, *Neural Computing and Applications*, vol. 28, pp. 1–11, 2015.

266. L. Wang, M. Duan, and S. Duan, Memristive perceptron for combinational logic classification, *Mathematical Problems in Engineering*, vol. 2013, pp. 1–7, 2013.

267. A. Galushkin, Neural networks realizations using memristors, in 2014 International Conference on Engineering and Telecommunication (EnT). IEEE, 2014, pp. 77–81.

268. M. S.Feali and A. Ahmadi, Transient response characteristic of memristor circuits and biological-like current spikes, *Neural Computing and Applications*, pp. 1–11, 2016.

269. E. Gale, B. de Lacy Costello, and A. Adamatzky, Emergent spiking in non-ideal memristor networks, *Microelectronics Journal*, vol. 45, no. 11, pp. 1401–1415, 2014.

270. E. Gale, B. de Lacy Costello, and B.Adamatsky,, Spiking in memristor networks, inA. Adamatsky and L. Chua (eds), *Memristor Networks*. Springer, 2014, pp. 365–387.

271. M. S.Feali and A. Ahmadi, Realistic Hodgkin-Huxley axons using stochastic behavior of memristors, *Neural Processing Letters*, pp. 1–14, 2016

272. M. D. Pickett, G. Medeiros-Ribeiro, and R. S. Williams, A scalable neuristor built with Mott memristors, *Nature Materials*, vol. 12, no. 2, pp. 114–117, 2013.

273. A. Cassidy and A. Andreou, Beyond Amdahl's law: An objective function that links multiprocessor performance gains to delay and energy. *IEEE Transaction on Computures.* vol. 61, pp. 1110–1126, 2012.

274. R. Brette and D. Goodman, Simulating spiking neural networks on GPU Netw. *Compuation in Neural System.* vol. 23, pp. 167–182, 2012.

275. D. Neil and S.-C. Liu, Minitaur, an event-driven FPGA based spiking network accelerator *IEEE Transaction on VLSI System*. vol. 22, pp. 2621–2628, 2014.

276. C.Farabetet al., Neuow: a runtime reconfigurable data flow processor for vision, Proc. IEEE Computer Society Conf. on Computer Vision and Pattern Recognition, pp. 109–116, 2011.

277. A. Majumdar, S. Cadambi, M. Becchi, S.Chakradhar and H. Graf, A massively parallel energy efficient programmable accelerator for learning and classification *ACM Trans. Architect. Code Optim*, vol. 9, no. 6, 2012.

278. S. Furber, F. Galluppi, S. Temple and L. Plana, The SpiNNaker project Proc. IEEE, vol. 102, pp. 652–665, 2014.

279. P. A. Merolla et al., A million spiking-neuron integrated circuit with a scalable communication network and interface, *Science* vol. 345, pp. 668–673, 2014.

280. B. V. Benjaminet al., Neurogrid: A mixed-analog-digital multichip system for large-scale neural simulations, Proc. IEEE, vol. 102, pp. 699–716, 2014.

281. J. Schemmel, D. Bruderle, K. Meier and B. Ostendorf, Modeling synaptic plasticity within networks of highly accelerated I& F neurons, *Proc. IEEE Int. Symp. Circuits and Systems* pp. 3367–3370, 2007.

282. N.Qiaoet al.,A re-configurable on-line learning spiking neuromorphic processor comprising 256 neurons and 128 k synapses, *Frontiers Neurosci*, vol. 9, p. 141, 2015.

283. B. Zhang, L. Shi and S.Song, Creating more intelligent robots through brain inspired computing, special supplement: Brain inspired robotics, *Science*, vol. 354, pp. 4–9, 2016.

284. S. Gupta and M. Rajasekhara Babu. Performance analysis of GPU compared to single core and multi-core CPU for natural language applications *Int. J. Adv. Comput. Sci. Appl*. vol. 2, pp.50–53, 2011.

285. T. Luo et al., DaDianNao: A neural network supercomputer. *IEEE Trans. Comput.*, vol. 66, pp. 73–88, 2017.

286. http//isscc:org=wp//content//uploads//sites//10//2017//05//ISSCC2017P ressKit:pdf.

287. T. M. Taha, R. Hasan, C. Yakopcic, and M. R. McLean, Exploring the design space of specialized multicore neural processors, Proc. Int. Joint Conf. on Neural Networks (IJCNN), 2013.

288. W.-H.Chenet al., A 65 nm 1 Mb nonvolatile computing-in-memory ReRAM macro with sub-16 ns multiply-andaccumulate for binary DNN AI edge processor, ISSCC Digest of Technical Papers, pp. 494–496, 2018.

289. H.-S. Lee, S. Park and D.-H. Lee, RMSS: An efficient recovery management scheme on NAND ash memory based solid state disk, *IEEE Trans. Consum. Electron*.vol. 59, pp.107–112, 2013.

290. M. Nigus, R. Priyadarshini & R.M. Mehra, Stochastic and novel generic scalable window function-based deterministic memristor SPICE model comparison and implementation for synaptic circuit design. *SN Appl. Sci.* vol. 2, p. 128. https://doi.org/10.1007/s42452-019-1888-z2020

291. G. W. Burr, R. M. Shelby, A. Sebastian et al., Neuromorphic computing using non-volatile memory, *Advances in Physics: X*, vol. 22, no. 1, pp. 89–124, 2017.

292. D. Soudry, C.D. Di, A. Gal et al.,Memristor-based multilayer neural networks with online gradient descent training. *IEEE Transactions on Neural Networks & Learning Systems*, vol. 26, no. 10, p. 2408, 2015.

293. E. Rosenthal, S. Greshnikov, D. Soudry et al., A fully analog memristor-based neural network with online gradient training. InIEEE International Symposium on Circuits and Systems. IEEE, pp. 1394–1397, 2016.

294. J. Chung, J. Park, and S. Ghosh, Domain wall memory based convolutional neural networks for bit-width extend ability and energy-efficiency, in International Symposium on LowPower Electronics and Design. ACM, pp. 332–337. Memristor Neural Network Design http://dx.doi.org/10.5772/intechopen.69929 279, 2016.

295. I. Kataeva, F. Merrikh-Bayat, E. Zamanidoost et al., Efficient training algorithms for neural networks based on memristive crossbar circuits, in *IEEE International Joint Conferenceon Neural Networks*, pp. 1–8, 2015.

296. P. Y.Simard, D.Steinkraus, J. C.Platt. Best practices for convolutional neural networks applied to visual document analysis, in International Conference on Document Analysis & Recognition (ICDAR), vol. 3, pp. 958–963, 2003.

297. A. R.Mohamed, G. E.Dahl, G. Hinton, Acoustic modeling using deep belief networks. *IEEE Transactions on Audio, Speech, and Language Processing*, vol. 20, no.1, pp. 14–22, 2012.

298. T. Mikolov, M. Karaat, and L. Burget et al., Recurrent neural network based language model, inInterspeech, Conference of the International Speech Communication Association, Makuhari, Chiba, Japan, pp. 1045–1048, 2010.

299. S. G. Hu, Y. Liu, Z. Liu et al., Associative memory realized by a reconfigurable memristive Hopfield neural network. *Nature Communications*, vol. 6, p. 7522, 2015.

300. M.Nigus, R.Priyadarshin, and R.M. Mehra, Memristive biophysical neuron models forming an excitatory-inhibitory neural network for modeling PING rhythm generation, *Journal of Computational Electronics*, vol. 20, pp. 681–708, 2021. https://doi.org/10.1007/s10825-020-01580-9

301. M. A.Zidan, A.Chen, G.Indiveri, and W. D.Lu, Memristive computing devices and applications, *Journal of Electroceramics*, vol. 39, no.1, pp. 4–20, 2017. https://doi.org/10.1007/s1083 2-017-0103-0

302. K. A.Campbell,Self-directed channel memristor for high temperature operation, *Microelectron J.*, vol. 59, pp. 10–14, 2017.

303. Anping Huang, Xinjiang Zhang, Runmiao Li, and Yu Chi, Memristor Neural Network Design, Memristor and Memristive Neural Networks, Alex Pappachen James, IntechOpen, 2017 (December 20). DOI: 10.5772/intechopen.69929. Available from: www.intechopen.com/books/memris tor-and-memristive-neuralnetworks/ memristor-neural-network-design.

304. B. Liu, Memristor-based analog neuromorphic computing engine design and robust training scheme [Dissertation]. University of Pittsburgh; 2014.

305. Y.Zhang, Z. Wang, J. Zhu, Y. Yang, M. Rao, W.Song, Y. Zhuo, X. Zhang, M. Cui, L. Shen, and R. Huang, Brain-inspired computing with memristors: Challenges in devices, circuits, and systems. *Applied Physics Reviews*, vol.7, no.1, p. 011308, 2020.

306. H. Jeong, and L. Shi, Memristor devices for neural networks, *Journal of Physics D: Applied Physics*, vol. 52, no.2, p.023003, 2018.

307. L. Shi, G. Zheng, B. Tian, B. Dkhil, and C. Duan, Research Progress on Solutions to the Sneak Path Issue in Memristor Crossbar Arrays. 2020. https://doi.org/10.1039/D0NA00100G

308. C.D. Schuman, T.E. Potok, R.M.Patton, J.D. Birdwell, M.E. Dean, G. S. Rose, and J.S. Plank, A survey of neuromorphic computing and neural networks in hardware. *arXiv preprint arXiv:1705.06963*, 2017.

# CHAPTER 11

# Machine Learning Applications to Recognize Autism and Alzheimer's Disease

Touko Tcheutou Stephane Borel
and Rashmi Priyadarshini

*Sharda University, Greater Noida, Uttar Pradesh, India*

*Email id: stephtouko@yahoo.com; rashmi.priyadarshini@sharda.
ac.in*

## 11.1 INTRODUCTION

Autism spectrum disorder (ASD) is one of a cluster of cerebral diseases involving impairments in communication/social interactive skills, mood, attention, cognitive and adaptive skills, and cognitive functions. That is, a set of neurodevelopmental impairments causing difficulty in connecting with other people. ASD is characterized by repetitive, cyclic, and obstructive behaviors, with symptoms stemmingfrom a convolutedgenotype–phenotyperelationship wherein pre-existing neurodevelopmental liabilities interact with the child's environment. In responsive modification, the child typically develops compensatory tactics and defense mechanisms. Studies of children at high genetic ASD risk, defined by an older diagnosed sibling, are discovering developmental corridors to phenotype manifestation [1]. The severity of ASD is greater in terms of social impact, rather than morbidity. Autistic behavior is influenced by the environment and genesbut the origins of the brain disorder remain a mystery. It is thought that there could be an interaction between environmentalfactors and the genes of the patient, which could cause certain deformations in the operative connectivity and cerebral development [1]. Children on the autism spectrum display concurrentsense-processing complications and are clinically treated using self-modifying mediation. Contemporary therapy utilizes sensory interventions usingvarious hypothetical paradigms, which have differing goals, and deploy a multiplicity of sensory modalities consisting of remarkably disparate procedures. Earlier evaluations studied the effects of sensory interventions without recognizing such empirical contradictions [2]. ASD diagnoses are typically delayed, resulting in unrealized treatment opportunities during the formative period of development. Our investigation extrapolates previous assessments of age-related factors in ASD diagnosis, offering clinical research recommendations, programs, and early detection methods [3]. Mutations affect typical neurodevelopment in utero through to adolescence via gene composites involved in exuberant synaptogenesis and axon motility.

Important knowledge about brain deformation in people with ASD has emerged thanks to recent advances in the science of neuroimaging. Amygdala and nucleus are the two most important elements in ASD, which is why ASD is prevalent in neuroimaging and neuropathological studies [4]. ASD can be prevented or treated in both men and women, but according to diagnostics, the rate is higher in men than in women [5].

Studies of heredity have unearthed hundreds of gene deviations in autism with radically differing risk effects habitually associated with similar conditions. However, numerous variations coalesce into mutual biological pathways, indicating characteristically pervasive autism traits, including logical heterogeneity, variablepenetrance, and genetic pleiotropy [5].

## 11.2 BRAIN DISORDERS

### 11.2.1 Autism Spectrum Disorder (ASD)

Researchers believe that typical ASD symptoms in children should be considered as ASD, while adult symptoms should not. Few behavioral indicators are diagnosed in the child's first year, with most emerging in the second [1]. According to the *Diagnostic and Statistical Manual of Mental Disorders*, 5th Edition (DSM-5), ASD sufferers react inappropriately to conversational cues and engage in abnormal routines and inappropriate obsessions, and can struggle with relationships [6]. ASD patients present the gamut of cognitive aptitudes, from acute intellectual retardation to remarkable intelligence. The DSM-5 does not include postponement of lingual acquisition as a core ASD symptom, as not all ASD sufferers display this trait. The eponymous condition is considered the most acute amongautism spectrum disorders and distinguishes it from disorders such as Asperger's syndrome; the percentage of children with ASD is approximately 0.2% (20 out of 10,000). Research using functional magnetic resonance imaging (FMRI) on the brains of people with ASD suggests a substantial decrease in long-distance connectivity. Adaptation of cell division, elevated inflammation of neurons and apoptosis are the factors that cause perturbations in the brain and prevent it from developing in the usual way. Recent studies have observed hypo-connectivity and hyper-connectivity issues in autistic children's brains, depending on age-related factors, for example a child of three months at high risk of developing autism displays elevated connectivity in contrast to low-risk children, a variant which dwindles between the ages of six and nine months. According to research, malformations (morphologies), deterioration of brain organs, amygdale are the different pathologies that the autistic brain suffers. At six months the cephalic circumference of ASD infants accretes rapidly when compared to typical infants, but declines at adolescence, resulting in a typical adult cerebral mass and volume. Mounting evidence suggests the significance of mirror neuronbrain cells activated when an individual executes and observes a motor action. Motor neurons attack much more on the recognition of others, emotions and the regulation of cognitive tasks. Sympathy, regret, compassion, and so onareenabled in human beings by the mirror neuron system: this system also allows the processing of the visual involved in speech, the planning of movements, and memory. Several studies have shown that electroencephalography, FMRI and electromyography are the ranges of techniques that sufficiently demonstrate deficiencies in the mirror neuron system when children have ASD, which hinders their understanding and recognition of the motor activity of others.

The genetic origins of this disorder in neuron development have been shown to be difficult via whole exome sequencing (WES) and cytogenetics. Likewise, studies of twins and families suggest that ASD's heritability is more than 80%. The principal ASD- associated syndromes are fragile X syndrome (FXS) and tuberous sclerosis (TS), both of which have similar pathophysiological processes to those of ASD, including deviant mRNA translation and elevated synthesis of protein. FXS is a genetic disease caused by aninconsistent increase of the FMR1 gene's multiple CGG repeat and is characterized by abnormal facial features as well as variously severe cognitive deficiencies. TS presents asepilepsy, learning challenges, and social interactive issues. More than 40% of patients with TS also suffer from ASD, hence the elevated incidence of epileptic seizures in both ASD and TS sufferers.

WES and chromosomal microarray are the most common methods for identifying ASD-predisposition genes. WES recognizes new or rare genetic flaws in various heterogeneous disorders, such as ASD. A recent study of 928 patients showed that ASD is linked to intensely disruptive *de novo* in brain-expressed genes [7]. A fundamental component of ASD-related behavioral/functional performance is faulty sensory processing. Clinicians attribute repetitive behaviors, including twirling, swinging, and rotation, to sensory processing problems, noting that young people with ASD and other stereotypical phenomena suffered from much greater sensory processing problems (d=2) compared to controls. Rigidity, such as refusal to switch to a new activity or behavior, can be triggered by hyperactivity or hypo reactivity.

ASD-related sensory-processing issues may also affect children's diurnal functional performance, including eating, sleeping, and bath or bedtime behaviors. Selective eating in children is often accompanied by gustatory and/or olfactory over-sensitivities leading to specific food antipathies. Hyper reactivity and taste-aversion often cause anxiety/rigidity with regard to ingestion, evolving into anxious/disruptive eating behaviors. Sensory-processing issues may also upset the patient's sleep cycle, as sensory modulation issues in ASD children are linked to unstable sleeping patterns, specifically related to entering REM sleep, with 50%–80% of children with ASD afflicted. Additional studies are required to determine how sensory processing andhypo reactivity or hyper reactivity influence self-modulation and stimulation [2].

Autism is distinguished by: (i) a qualitative impairment in communication in society via non-verbal behaviors, eye contact, gestures and physical behavior, which will lead to a lack of connection and spontaneity, sharing of interests or social reciprocity; (ii) qualitative impairments in social communication, which manifest as delayed language development lacking non-verbal compensation, difficulties in starting and maintaining conversation, stereotypical language, and a lack of creative invention or imagination; (iii) a repertoire of limited interest, an obsession with object or subject, slavery to a dysfunctional ritual/diet/routine, stereotypical motor mannerisms and fixation on parts or aspects of object instead of gestalt. Sensory aberrations are common, including hypo-/hypersensitivity and preoccupation with certain sensations. The lack of imaginative play implies problems with idea generation, which isessential for human bonding. Numerous factors have increased early referral and diagnosis of autism: (i) heightened detection by healthcare professionals of early autism symptoms, leading to earlier recommendation to pediatric and child-development experts; (ii) increased public and media interest involving the publication of memoirs and bio-graphical journalism, including the depiction of pediatric ASD behavior, leading to parental help outreach.

Screening methods applied to both referred and entire populations (CHAT which means Checklist for Autism in Toddlers) have identified autism as early as 18 months. However, in the only general population study to date, although CHAT screening presented an important positive predic-tion, its sensitivity rate was insignificant and cannot be recommended for populations digging at any given time.

Diagnosing autistic people at the age of two years is less accurate/stable than that of related ASD, but the preliminary response should not be impul-sive as working diagnoses are in most cases refined over time in conference with parents. Medical evaluations must identify difficulties in early non-verbal interaction characteristic of children with ASD from an age of two years onwards. The particular syndrome presenting in a two-year-old with ASD may vary widely from a more typicalfour- or five-year-old. In par-ticular, explicit repetitive and/or stereotyped behaviors may be present in a minority of cases, but when concordant with social/interactive deformity strongly indicate ASD [8].

The brain architecture in patients with ASD, both in the frontal and tem-poral lobes, is noticeably disturbed, especially the amygdala, which is influ-ential on cognition as evidenced by numerous neuro studies (pathological and imaging). The median temporal lobe prior to hippocampal construc-tion is essential for declarative memory (conscious memory of facts and events) and determines antisocial/social behavior in patients with ASD. The amygdala is the core of the limbic system and determining eye contact and facial recognition/motion.

Injury to the amygdala is manifested by changes in the treatment of fear, modulation of memory with emotional content and eye contact with the human face. The effects in people with tonsil lesions mimic the phenomena of ASD because the amygdala processes sensory, audio-visual, visceral and synesthetic inputs, channeling through the main efferent ducts, striaterminalis. The amygdala comprises 13 nuclei that, historically and chemically, are divided into three groups: centro medial (CM), basolateral (BL) and superficial groups. The BL group has a functioning comparable to that of a sensory stimulus that connects nodes for large social cognition, connecting the superficial groups and the CM with reciprocity to the orbitofrontal, medial prefrontal (mPFC) and anterior cingulate (ACC) cortex.

Neurochemical investigations of the amygdala have revealed an elevated opiate and benzodiazepine, a receptor density including cholinergic, dopaminergic, noradrenergic, and serotonergic cell bodies and pathways. Because a population of aggressive patients with temporal epilepsy experienced a decrease in aggression after bilateral stereotactic removal of the tonsilloid and corticomedia nuclei, the function of the amygdala in emotional processing, particularlyrabies, was studied with confirmation of tonsil deficiency in patients with ASD. Post-mortem investigations have revealed higher incidence of amygdala disease in ASD patients in comparison, by ageandsex, to control matching.

Finally, studies tracking the younger siblings of ASD patients suggest investigative prejudice causes an overemphasis on gender-bias, particularly in the high functioning control. Self-advocates have declared timely and precise diagnoses an 'essential need'. Despite diagnostic variance, the gender disparity in ASD predominance remains at 2:1–3:1, indicating the need for the investigation of sexual dimorphism in the symptomatology of ASD. The 'female protective effect' (FPE) suggests that ASD females are at less risk of certain ASD symptoms than affected males. The FPE has, in other disorders, such as clubfoot, been ascribed to a clear gender prejudice, and genetic research of ASD cohorts has discovered a greater liability for *de novo* copy number variation (CNV) and *de novo* loss of function-point aberration in ASD females than in male counterparts. In addition, inherited small CNV are transmitted more often from unafflicted mothers than counterpart fathers [5].The clinical benefits of early detection imply the importance of parental intervention for autistic children. Heightened parental intervention dates back three decades, encouraging learning in ASD children. Parents trained as 'co-therapists' with consistent interventional handling and modification of the child's ASD behaviors enhance formative pediatric social interactions, including heightened skills and confidence, in addition to lessening parental/filial stress. Group parental skill coaching has been proven to promote communal support. The metrics of parental intervention in pediatric development must include: pediatric developmental improvement, parental–filial communication, parental understanding, outlook and anxiety, familial cohesion, and cost benefit analysis. The majority of assessments that involve timely parental intercession have not been systematic, thus reducing their scope and validity as they include uncontrolled studies that rely on single case studies without clinical precision, applicable universality, methodological compliance, often omitting statistics that did not benefit their prognosis. Case-Smith employed a restrictive foundation for result comparison by favoring children's cerebral performance, while knowing that the majority of autistic children have issues withmental retardation [2]. The New York State Department of Health performed the most responsible clinical study of timely autism intervention, allegedly to develop scientific proceduresby law, but flagrantly failed to investigate parental intervention. Other self-motivated clinicians made ostensibly methodical studies of randomized controlled trials of parent arbitered timely intervention by revealing only child-related results, knowing that parental intervention would hurt their medical practice by removing their own importance [9,10].

In an attempt to understand the genetic systems engendering ASD risk, researchers have studied various afflicted families, including those with con-sanguinity and single (simplex family) and multi-affected family members withASD, often spanning numerous generations [11–13].

## 11.2.2 Alzheimer's Disease (AD)

Alzheimer's disease (AD) is considered to be a progressive and irreversible disorder that can cause thought and memory loss due to the slow deterioration of brain cells. Neurodegenerative dementia is a progressive deterioration of this disease. Several factors come into play when diagnosing AD, including: the patient's history, a mental state examination, and a neurobiological examination [14–15]. Patients remain immobilized when FMRI analyses are required, which allows for good data acquisition of data on how the brain works [16–18]. Low signal strength can be seen in different MRI and fMRI techniques [19–21], when brain cells have degenerated. When a person is already elderly, it is very difficult to distinguish the visual identification of the AD from the images of the patient. This requires a great deal of knowledge in the field [14]. Clinicians have always been interested in the development of an algorithm that will allow them to classify image data on MR and distinguish brain diseases from healthy patients [14–22].

### 11.2.3 Mild Cognitive Impairment

Cognitive impairment and related conditions have been studied in the field of neuroscience for several years. Someresearchers have proposed that mild cognitive impairment should be considered as an early state of cognitive impairment but still abnormal. Since then, mild cognitive impairment has been studied extensively in clinical and research projects. Many epidemiological research projects have shown an accelerated and progressive rate to dementia and AD in mild cognitive impairment patients. However, a broadly constant definition of mild cognitive impairment that is acceptable for researchers has not been obtained yet due to the controversy over the concept of mild cognitive impairment. Scientists have proposed that the diagnosis of mild cognitive impairment should be performed in the same fashion as AD. Machine learning algorithms (discussed below) have been developed to identify subjects and classify diseases into the various types of mild cognitive impairment [23–25].

## 11.3 DEEP LEARNING

Deep learning is a modern method of normal learning that has been designed by the human brain. Thanks to the complex algorithms, this technique could be realized, making it possible to model and extract data from a similar architecture of much more complicated neurons. The 'neocortex', which is an integral part of the cerebral cortex, has been discovered by scientists and is concerned with hearing and eyes in mammals. It processes the signals of meaning and has served to develop deep learning by focusing on computer algorithms [22–26]. The classical neural networks are scientifically similar to convolutional neural networks (CNNs). Several parameters are reduced when it comes to setting up the topology of CNNs and thus allowing training to the back propagation [7–30].

In CNN, small portions of the image (called local receptive fields) are treated as inputs to the lowest layer of the hierarchical structure. The treatment of local receptive fields is like that of the hierarchic structure in the CNN. The level of invariance at offset is one of the main characteristics of CNN because the receptive field allows neurons to access oriented corners and edges. The standardization and pool layers each perform at their respective levels, one application of the activation function per

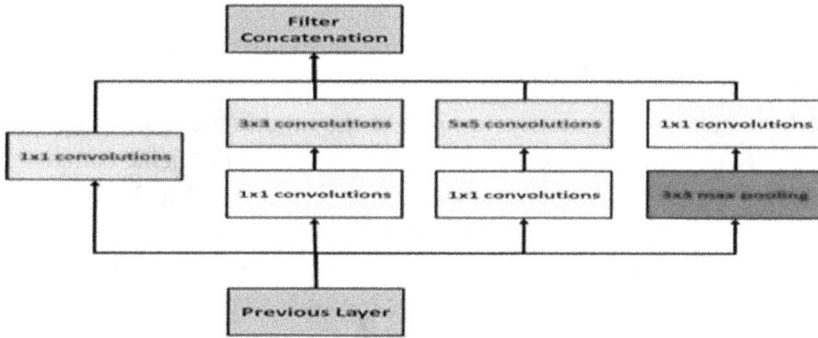

FIGURE 11.1    Reduction in CNN architecture.

element and one sampling operation. All this does not make any changes to the size of the image [26–27, 31]. Class scores are calculated by the fully connected layer from where a large number of classes come.

In CNN architecture, the convoluted layer can accept any volume of image by using this equation: $W_1 \times H_1 \times D_1$ where: F: spatialextent

S: size of stride

The convoluted layer outputs o f the new image are: $W_2 \times H_2 \times D_2$

It is important to note that to achieve the effect of filters, the way in which the layer produces the new output images is moreimportant than on the input images. Several CNN and more complex architectures have been made to recognize many objects derived from high-density data. By increasing the size of the layers, we can improve the CNN architecture.

## 11.3.1 ASD and Deep Learning

A new approach to deep learning has emerged and learns from transcription factor data and profiling, regulation code and what allows good prediction of nervous system working [32–38]. We demonstrate its value by focusing on the causal contribution of noncoding variants to clinical outcomes in ASD. Thanks to the causal participation of non-coding variants, we can demonstrate its value in the different clinical outcomes regarding ASD. To detect the stereotypes of the motor movement, we can use the proposed architecture with a 3-layer CNN.

To represent populations in a graph with imagery-based nodes and stop weight allows an assessment to be made on the effect of the disease and its components on each individual. A feature selection method of finding functional connectivity of regions of interest as predefined regions followed by using a denoising autoencoder was adopted. The extracted features were used to train and validate various machine learning models, such as support vector machines, random forest, and deep neural networks. However, the performance of classification in none of the models was promising. This might be explained by the feature extraction section not being completely successful [39]. An ensemble of a two-layer neural net that shares hidden nodes across tissues with a maximum of 30 hidden variables and also with sigmoidal non-linearity and SoftMax output was designed by Xiong et al. to predict RNA splicing levels from DNA sequences [40–43]. The segmentation of the brain into two parts (CWM, Cx) was proposed by the CAD system. The classification and control of the brains of people with autism provide a fusion of 8 extradite characteristics of the reconstructed meshes. Each mesh has a number of nodes equal to 48K and the respective average percentage for segmentation are: CWM (94.7%) and Cx (93.8%) [44–49].

## 11.3.2 Alzheimer's and Deep Learning

Many researchers and scientists have studied the question of brain function, including AD. The modeling of the prevention of this disease has been extensively studied in imaging. Among these researchers, we can mention, among others, Suk et al., who were able to study the classification of imaging data on deep learning, thus achieving the following results: MCI (95.9%), PET for the MCI converter (85%), structural MRI (75.8%). They were able to develop an encoder to extract features from low to medium level images [50–54].

Deep Alzheimer Disease is a deep learning based approach in which aggressive preprocessing steps against functional and structural MRI data were utilized, and enabled the CNN model to learn the patterns from the data more efficiently. To better distinguish between images of healthy patients [56–60] and patients with AD, several steps must take place, leading to the development of a pipeline. Among these steps we can mention: preprocessing, data conversion, and finally classification. Speaking of preprocessing, it can be used according to two methods: 4D rs-fMRI and 3D MRI structural. Following this step, comes the conversion of the image data in order to enter the classification based on deep learning, the architecture being ready to receive images in its input part based on CNN. Figure 11.2 shows in detail the whole process.

Figure 11.3 represents the abnormal brain having Alzheimer disease in the bottom right and other side the normal brain in the bottom left. We can identify the signal intensities of different brains.

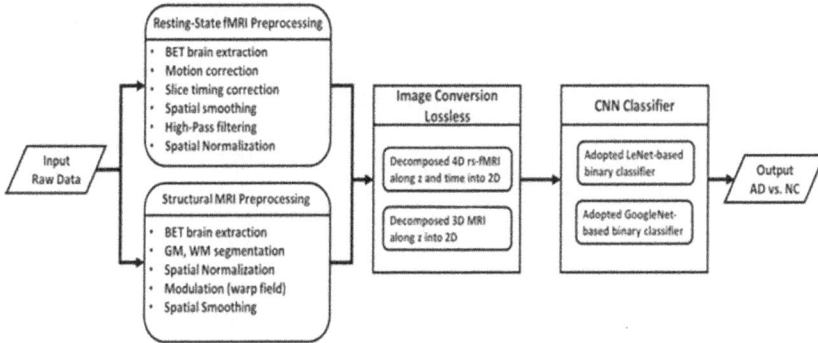

FIGURE 11.2    Classification methods based on CNN.

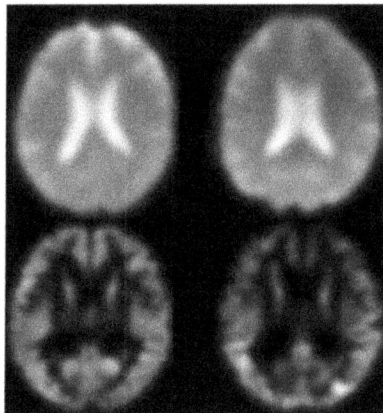

FIGURE 11.3    A middle cross-section of MRI data.

Figure 11.3 highlights the normal control of a brain; Figures 11.4 and 11.5 demonstrate how images are visualized by clinical scientists for fMRI and MRI models (20 filters each having a size of 5x5 pixels).

## 11.4 CONCLUSION

In this chapter the crucial brain disorders ASD and AD were discussed from various points of view, such as clinical symptoms and diagnosis through imaging techniques. As discussed, computational neuroscience and machine learning techniques have been recently utilized to solve certain prediction and classification tasks in the field of brain imaging. Although the developed techniques and pipelines based on machine learning are not

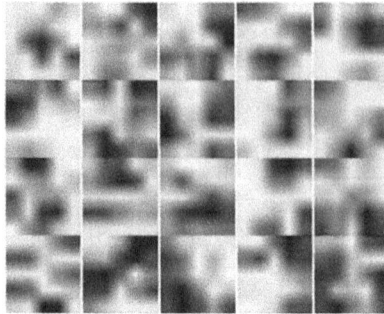

FIGURE 11.4    The first layer of LeNet in a fMRI model.

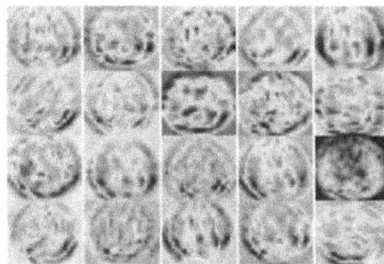

FIGURE 11.5    LeNet model.

able to predict the disorders with complete accuracy, the improvement in accuracy rates for these methods are promising and significant. However, such systems should not replace scientistsandspecialists in the field but rather enable them to make more reliable decisions in recognizing these brain disorders in the early stages.

## REFERENCES

1.  Jones, E. J. et al. 2014. Developmental pathways to autism: a review of prospective studies of infants at risk. *Neurosci. Biobehav. Rev.* 391–33.
2.  Case-Smith J., Weaver, L. L. and Fristad, M. A. 2015. A systematic review of sensory processing interventions for children with autismspectrum disorders. *Autism* 19 133–48.
3.  Daniels, A. M. and Mandell, D. S. 2014 Explaining differences inageatautismspectrum disorder diagnosis: a critical review. *Autism* 18583–97.
4.  Park, H. R. et al. 2016. A short review on the current understanding of autism spectrumdisorders. *Exp. Neurobiol.* 251–13.

5.  Halladay, A. K. et al. 2015. Sex and gender differences in autism spectrum disorder: summarizing evidence gaps and identifying emerging areas of priority. *Mol. Autism* 636.

6.  American Psychiatric Association 2013. *Diagnostic and Statistical Manual of Mental Disorders (DSM-5)*, 5th edn (Arlington, VA: American Psychiatric Publishing).

7.  Fakhoury, M. 2015. Autistic spectrum disorders: a review of clinical features, theories and diagnosis. *Int. J. Dev. Neurosci.* 4370–7.

8.  Charman, T. and Baird, G. 2002. Practitioner review: diagnosis of autism spectrum disorder in2-and3-year-oldchildren. *J. Child Psychol. Psychiatry* 43289–305.

9.  Heinsfeld, A. S. et al. 2018. Identification of autism spectrum disorder using deep learning and the ABIDE dataset. *NeuroImage* 1716–23.

10. McConachie, H. and Diggle, T. 2007. Parent implemented early intervention for young children with autism spectrum disorder: a systematic review. *J. Eval. Clin. Pract.* 13120–9.

11. Vorstman, J. A. et al. 2017Autism genetics: opportunities and challenges for clinical translation. *Nat. Rev. Genet.* 18(6), pp. 362–376.

12. Chambless, D. L. and Hollon, S. D. 1998. Defining empirically supported therapies *J. Consult.Clin. Psychol.* 66 7.

13. Nathan, P. E. 2007. Efficacy, effectiveness, and the clinical utility of psychotherapy research. *The Art and Science of Psychotherapy*, 69–83.

14. Vemuri, P., Jones, D. T. and Jack, C. R. 2012. Restingstate functional MRI in Alzheimer's disease. *Alzheimer's Res. Ther.* 42.

15. He, Y. et al. 2007. Regional coherence changes in the early stages of Alzheimer's disease: a combined structural and resting-state functional MRI study. *NeuroImage* 35488–500.

16. Sarraf, S. and Sun, J. 2016. Advances in functional brain imaging: a comprehensive survey for engineers and physical scientists. *Int. J. Adv. Res.* 4640–60.

17. Grady, C. et al. 2016. Age differences in the functional interactions among the default, front parietal control, and dorsal attention networks. *Neurobiol. Aging* 41159–72.

18. Saverino, C. et al. 2016. The associative memory deficit in aging is related to reduced selectivityof brain activity during encoding. *J. Cogn. Neurosci.* 281331–44.

19. Warsi, M. A. 2012. The fractal nature and functional connectivity of brain function as measured by BOLD MRI in Alzheimer's disease. *PhD Dissertation.*

20. Grady, C. L. et al. 2003. Evidence from functional neuroimaging of a compensatory prefrontal network in Alzheimer's disease. *J. Neurosci.* 23986–93.

21. Grady, C. L. et al. 2001Altered brain functional connectivity and impaired short-term memory in Alzheimer's disease. *Brain* 124739–56.

22. Raventós, A. and Zaidi, M. 2015. Automating neurological disease diagnosis using structural MR brains can features. http://cs229.stanford.edu/proj2015/262_poster.pdf.

23. Petersen, R.C. 2004. Mild cognitive impairment as a diagnosticentity. *J Intern.Med.* 256, 183–94.

24. Mitchell, A. J.and Shiri-Feshki, M.2009. Rate of progression of mild cognitive impairment to dementia–meta-analysis of 41 robust inception cohortstudies. *ActaPsychiatr. Scand.* 119, 252–65.

25. Fitzpatrick-Lewis, D. et al. 2015. Treatment for mild cognitive impairment: a systematic review and meta-analysis. *CMAJ Open* 3E419.

26. LeCun, Y. et al. 1998. Gradient-based learning applied to document recognition. Proc. *IEEE* 86, 2278–324.

27. Jia, Y.et al. 2014. CAFFE: convolutional architecture for fast feature embedding. *Proc. of the 22nd ACM International Conference on Multimedia* (New York: ACM).

28. Ngiam, J.et al. 2011. Multimodal deep learning. *Proc. of the 28th International Conference on Machine Learning (ICML-11)*

29. Erhan, D. et al. 2010. Why does unsupervised pre-training help deep learning? *J.Mach. Learn. Res.* 11, 625–60.

30. Schmidhuber, J. 2015. Deep learning in neural networks: an overview. *Neural Netw.* 61, 85–117.

31. Szegedy, C. et al. 2015. Going deeper with convolutions. *Proc. of the IEEE Conference on Computer Vision and Pattern Recognition*

32. Wang, L. et al. 2015. Object-scene convolutional neural networks for event recognition in images. *Proc. of the IEEE Conference on Computer Vision and Pattern Recognition Workshops.*

33. Arel, I., Rose, D. C. and Karnowski, T. P. 2010. Deep machine learning: a new frontier in artificial intelligence research. *IEEE Comput. Intell. Mag.* 513–18.

34. Krizhevsky, A., Sutskever, I. and Hinton, G. E. 2012. Imagenet classification with deep convolutional neural networks. *NIPS 12 Proc. of the 25th Int. Conf. on Neural Information Processing Systems* 1, 1097–105.

35. Lowe, D. G. 2004. Distinctive image features from scale-invariant key points. *Int. J. Comput. Vis.* 6091–110.

36. Simonyan, K. and Zisserman, A. 2014. Very deep convolutional networks for large-scale image recognition, *arXiv:1409.1556*

37. He, K. et al. 2016. Deep residual learning for image recognition. *Proc. of the IEEE Conference on Computer Vision and PatternRecognition*

38. Zhou, J., Theesfeld, C. and Troyanskaya, O. 2018. Decoding the role of noncoding genome in neurological disease with deep learning. *Biol. Psychiatry* 83, S82.

39. Rad, N. M. et al. 2018. Deep learning for automatic stereotypical motor movement detection using wearablesensors in autism spectrum disorders. *SignalProcess.* 144, 180–91.

40. Rad, N. M. et al. 2017. Deep learning for automatic stereotypical motor movement detection using wearable sensors in autism spectrum disorders, arXiv: 1709.05956.

41. Kong, Y. et al. 2019. Classification of autism spectrum disorder by combining brainconnectivity and deep neural network classifier. *Neurocomputing* 32463–8.

42. Parisot, S. et al. 2018. Disease prediction using graph convolutional networks: application to autism spectrum disorder and Alzheimer's disease. *Med. Image Anal.* 48, 117–30.

43. Xiong, H. Y. et al. 2015. The human splicing code reveals new insights into the genetic determinants of disease. *Science* 347 1254806.

44. Ismail, M. et al. 2017. Anewdeep-learning approach for early detection of shape variations in autism using structural MRI. *2017 IEEE Int.Conf. on Image Processing (ICIP)* (Piscataway, NJ: IEEE).

45. Dvornek,N. C., Ventola,P. and Duncan,J. S.2018Combining phenotypic and resting- tate fMRI data for autism classification with recurrent neural networks. *2018IEEE 15th Int. Symp. on Biomedical Imaging (ISBI 2018)* (Piscataway, NJ: IEEE).

46. Li, X. et al. 2018. 2-channel convolutional 3D deep neural network (2CC3D) for fMRI analysis: ASD classification and feature learning. *2018 IEEE 15th Int. Symp. on Biomedical Imaging (ISBI 2018)* (Piscataway, NJ: IEEE).

47. Andrews, D. S. et al. 2018. Using pattern classification to identify brain imaging markers in autism spectrum disorder. *Biomarkers in Psychiatry* (Berlin: Springer).

48. Guo, X.et al. 2017. Diagnosingautismspectrumdisorderfrombrainresting-statefunctional connectivity patterns using a deep neural network with a novel feature selection method. *Front. Neurosci.* 114, 60.

49. Jin, Y. et al. 2015. Identification of infants at high-risk for autism spectrum disorder using multi parameter multiscale white matter connectivity networks. *Hum. Brain Mapp.* 36, 4880–96.

50. Suk, H. I. and Shen, D. 2013. Deep learning-based feature representation for AD/MCI classification. *Int. Conf. on Medical Image Computing andComputer-Assisted Intervention* (Berlin: Springer).

51. Suk, H. I. et al. 2015. Latent feature representation with stacked auto-encoder for AD/MCI diagnosis. *Brain Struct. Funct.* 220, 841–59.

52. Suk H. I., Shen, D. and Initiative A. S. D. N. 2015. Deep learning in diagnosis of brain disorders. *Recent Progress in Brain and Cognitive Engineering* (Berlin: Springer) pp. 203–13.

53. Suk, H. I. et al. 2014. Hierarchical feature representation and multi-modal fusionwithdeep learning for AD/MCI diagnosis. *NeuroImage* 101, 569–82.

54. Payan, A. and Montana, G. 2015. Predicting Alzheimer's disease: a neuroimaging study with 3D convolutional neural networks, *arXiv:1502.02506*

55. Liu, S. et al. 2015. Multimodal neuroimaging feature learning for multiclass diagnosis of Alzheimer's disease. *IEEE Trans. Biomed. Eng.* 621, 132–40.

56. Arvesen, E. 2015. Automatic classification of Alzheimer's disease from structural MRI. *MSc Thesis.*

57. Liu, F. and Shen, C. 2014. Learning deep convolutional features for MRI based Alzheimer's disease classification, *arXiv: 1404.3366.*

58. Liu, S. et al. 2014. High-level feature based PET image retrieval with deep learning architecture. *J. Nucl. Med.* 55, 2028.

59. Brosch, T. Tam, R. and Initiative A. S. D. N. 2013. Manifold learning of brain MRIs by deep learning. *Int. Conf. on Medical Image Computing and Computer-Assisted Intervention* (Berlin: Springer).

60. Rampasek, L. and Goldenberg, A. 2016. TensorFlow: biology's gateway to deeplearning? *Cell Syst.* 2 12–14.

61. de Brebisson, A. and Montana, G. 2015. Deep neural networks for anatomical brain segmentation. *Proc. of the IEEE Conf. on Computer Vision and Pattern Recognition Workshops.*

62. Ijjina.E. P. and Mohan, C. K. 2016. Hybrid deep neural network model for human action recognition. *Appl. Soft Comput.* 46936–52.

63. Zhu, X., Suk, H.-I. and Shen, D. 2014. A novel matrix-similarity based loss function for joint regression and classification in AD diagnosis. *NeuroImage* 100 91–105.

64. Zu, C. et al. 2016. Label-aligned multi-task feature learning for multimodal classification of Alzheimer's disease and mild cognitive impairment. *Brain Imaging Behav.* 10, 1148–59.

65. Liu, M. et al. 2016. Inherent structure-based multiview learning with multitemplate feature representation for Alzheimer's disease diagnosis. *IEEE Trans. Biomed. Eng.* 63, 1473–82.

66. Li, F. et al. 2015. A robust deep model for improved classification of AD/MCIpatients. *IEEE J. Biomed. Health Inform.* 19, 1610–16.

67. Hosseini-Asl, E., Keynto, R. and El-Baz, A. 2016. Alzheimer's disease diagnostics by adaptation of 3D convolutional network, *arXiv:1607.00455*

68. Sarraf, S. and Tofighi, G. 2016. Deep learning-based pipeline to recognize Alzheimer's disease using fMRI data. *2016 Future Technologies Conf. (FTC)*

69. Sarraf, S. et al. 2017. *DeepAD: Alzheimer's Disease Classification via Deep Convolutional Neural Networks using MRI and fMRI* (bioRxiv). https://doi.org/10.1101/070441

# Index

For Product Safety Concerns and Information please contact our EU
representative GPSR@taylorandfrancis.com
Taylor & Francis Verlag GmbH, Kaufingerstraße 24, 80331 München, Germany

www.ingramcontent.com/pod-product-compliance
Lightning Source LLC
Chambersburg PA
CBHW060339220326
41598CB00023B/2752